Lecture Notes in Artificial Intelligence 13514

Subseries of Lecture Notes in Computer Science

Series Editors

Randy Goebel
University of Alberta, Edmonton, Canada

Wolfgang Wahlster
DFKI, Berlin, Germany

Zhi-Hua Zhou
Nanjing University, Nanjing, China

Founding Editor

Jörg Siekmann
DFKI and Saarland University, Saarbrücken, Germany

More information about this subseries at https://link.springer.com/bookseries/1244

Oscar Corcho · Laura Hollink · Oliver Kutz ·
Nicolas Troquard · Fajar J. Ekaputra (Eds.)

Knowledge Engineering and Knowledge Management

23rd International Conference, EKAW 2022
Bolzano, Italy, September 26–29, 2022
Proceedings

Springer

Editors
Oscar Corcho (ORCID)
Univ. Politecnica de Madrid
Boadilla del Monte, Spain

Laura Hollink (ORCID)
CWI
Amsterdam, The Netherlands

Oliver Kutz (ORCID)
Free University of Bozen-Bolzano
Bolzano, Italy

Nicolas Troquard (ORCID)
Free University of Bozen-Bolzano
Bolzano, Italy

Fajar J. Ekaputra (ORCID)
TU Wien
Vienna, Austria

ISSN 0302-9743   ISSN 1611-3349 (electronic)
Lecture Notes in Artificial Intelligence
ISBN 978-3-031-17104-8   ISBN 978-3-031-17105-5 (eBook)
https://doi.org/10.1007/978-3-031-17105-5

LNCS Sublibrary: SL7 – Artificial Intelligence

This Springer imprint is published by the registered company Springer Nature Switzerland AG
The registered company address is: Gewerbestrasse 11, 6330 Cham, Switzerland

Preface

This volume contains the proceedings of the 23rd International Conference on Knowledge Engineering and Knowledge Management (EKAW 2022), held in Bozen-Bolzano, Italy, during September 26–29, 2022.

We invited three types of papers: research papers, in-use papers, and position papers. Each could be submitted as a short paper with a maximum seven pages or as a long paper with a maximum of 15 pages.

Overall, we received 58 abstract submissions, of which 57 were eventually accompanied by a full paper, which were reviewed by 59 reviewers and 10 subreviewers. The review process was single-blind, i.e., the authors were known to the reviewers, while the reviewers remained anonymous to the authors. Each paper received three to four reviews, and discussions were encouraged by both program chairs for papers that exhibited strongly divergent opinions. In total, 16 papers were accepted for publication in this volume, of which 11 are full length research papers, three are short research papers, and two are short position papers.

The previous event in the series, EKAW 2020, introduced a special theme related to "Ethical and Trustworthy Knowledge Engineering." This theme is still very relevant in 2022, and thus has remained one of the core topics of the conference.

Sixteen papers were presented at the conference. In addition to paper presentations, the conference featured a Poster and Demo session, three keynote speeches, one workshop, and one tutorial. The Knowledge Management for Law (KM4LAW) workshop was organized by Davide Audrito (University of Bologna), Luigi di Caro (University of Turin), Francesca Grasso (University of Turin), Roberto Nai (University of Turin), and Emilio Sulis (University of Turin). A tutorial on "Trends in Terminology Generation and Modelling" was organized by Rute Costa (Universidade NOVA de Lisboa), Elena Montiel-Ponsoda (Universidad Politécnica de Madrid), Sara Carvalho (University of Aveiro), and Patricia Martín-Chozas (Universidad Politécnica de Madrid).

We would like to express our gratitude towards the Organizing Committee and the Program Committee. A specific thanks goes to the "emergency reviewers" who provided additional reviews within a short time span in cases where previously provided reviews did not lead to a clear decision to reject or accept a paper.

We would also like to thank our keynote speakers (Fabien Gandon, Hannah Bast, and Vanessa López) for accepting our invitations without hesitation and bringing their insights into the importance of knowledge engineering in today's world.

Finally, our gratitude goes also to the sponsors of the conference, the Free University of Bozen-Bolzano and Universidad Politécnica de Madrid, the Artificial Intelligence Journal, and to the local organization team for making it possible to have a physical event with so many activities focused on the networking of participants after these years of less social contact.

August 2022

Oscar Corcho
Laura Hollink
Oliver Kutz
Nicolas Troquard
Fajar J. Ekaputra

Organization

General Chairs

Oliver Kutz   Free University of Bozen-Bolzano, Italy
Nicolas Troquard   Free University of Bozen-Bolzano, Italy

Program Committee Chairs

Oscar Corcho   Universidad Politécnica de Madrid, Spain
Laura Hollink   Centrum Wiskunde & Informatica,
     The Netherlands

Posters and Demos Chairs

Danai Symeonidou   INRAE, Montpellier, France
Ran Yu   University of Bonn, Germany

Workshops and Tutorials Chairs

Davide Ceolin   Centrum Wiskunde & Informatica,
     The Netherlands
María Poveda   Universidad Politécnica de Madrid, Spain

Publicity Chairs

Claudenir M. Fonseca   Free University of Bozen-Bolzano, Italy
Guendalina Righetti   Free University of Bozen-Bolzano, Italy

Proceedings Chair

Fajar J. Ekaputra   TU Wien, Austria

Program Committee

Nathalie Aussenac-Gilles   IRIT and CNRS, France
Eva Blomqvist   Linköping University, Sweden
Stefano Borgo   ISTC-CNR, Italy
Philipp Cimiano   Bielefeld University, Germany
Miguel Couceiro   Inria, France

Claudia d'Amato   University of Bari, Italy
Mathieu D'Aquin   Loria, University of Lorraine, France
Enrico Daga   The Open University, UK
Jérôme David   Inria, France
Daniele Dell'Aglio   Aalborg University, Denmark
Sylvie Despres   LIM & BIO, Université Sorbonne Paris Nord,
                                       France
Anastasia Dimou   KU Leuven, Belgium
Mauro Dragoni   FBK-ICT Irst, Italy
Paola Espinoza   Universidad Politécnica de Madrid, Spain
Catherine Faron Zucker   Université Côte d'Azur, France
Jesualdo Tomás Fernández-Breis   Universidad de Murcia, Spain
Pablo Fillottrani   Universidad Nacional del Sur, Argentina
Fabien Gandon   Inria, France
Raúl García-Castro   Universidad Politécnica de Madrid, Spain
Daniel Garijo   Universidad Politécnica de Madrid, Spain
Giancarlo Guizzardi   Free University of Bozen-Bolzano, Italy, and
                                       University of Twente, The Netherlands
Torsten Hahmann   University of Maine, USA
Harry Halpin   World Wide Web Consortium, USA
Rinke Hoekstra   University of Amsterdam, The Netherlands
Antoine Isaac   Europeana and VU University Amsterdam,
                                       The Netherlands
Krzysztof Janowicz   University of California, Santa Barbara, USA
Clement Jonquet   LIRMM, University of Montpellier, France
Manolis Koubarakis   National and Kapodistrian University of Athens,
                                       Greece
Francesco Kriegel   TU Dresden, Germany
Adila A. Krisnadhi   Universitas Indonesia, Indonesia
Markus Krötzsch   TU Dresden, Germany
Agnieszka Lawrynowicz   Poznan University of Technology, Poland
Danh Le Phuoc   TU Berlin, Germany
Maxime Lefrançois   MINES Saint-Étienne, France
Fabrizio Maria Maggi   Free University of Bozen-Bolzano, Italy
Suvodeep Mazumdar   University of Sheffield, UK
Alessandro Mosca   Free University of Bozen-Bolzano, Italy
Till Mossakowski   University of Magdeburg, Germany
Enrico Motta   The Open University, UK
Vit Novacek   National University of Ireland, Galway, Ireland
Francesco Osborne   The Open University, UK
Heiko Paulheim   University of Mannheim, Germany
Yannick Prié   LINA, University of Nantes, France

Ulrich Reimer	Eastern Switzerland University of Applied Sciences, Switzerland
Oscar Rodríguez	Teach on Mars, France
Marta Sabou	WU Wien, Austria
Stefan Schlobach	Vrije Universiteit Amsterdam, The Netherlands
Barış Sertkaya	Frankfurt University of Applied Sciences, Germany
Cogan Shimizu	Kansas State University, USA
Armando Stellato	Tor Vergata University of Rome, Italy
Mari Carmen Suárez-Figueroa	Universidad Politécnica de Madrid, Spain
Vojtěch Svátek	Prague University of Economics and Business, Czech Republic
Annette Ten Teije	Vrije Universiteit Amsterdam, The Netherlands
Andrea Tettamanzi	Côte d'Azur University, France
Ilaria Tiddi	Vrije Universiteit Amsterdam, The Netherlands
Konstantin Todorov	LIRMM, University of Montpellier, France
Guohui Xiao	Free University of Bozen-Bolzano, Italy
Fouad Zablith	American University of Beirut, Lebanon
Ondřej Zamazal	Prague University of Economics and Business, Czech Republic

Additional Reviewers

Sergio Alejandro Gomez
Inès Blin
Victor Charpenay
Victor de Boer
Fatma-Zohra Hannou
Jan-Christoph Kalo
Pierre Monnin
Paula Reyero Lobo
Giuseppe Rizzo
Miroslav Vacura

Contents

Research Papers

Basic Human Values and Moral Foundations Theory in ValueNet Ontology

Stefano De Giorgis[1]([✉])(ID), Aldo Gangemi[1,2](ID), and Rossana Damiano[3](ID)

[1] University of Bologna, Via Zamboni 32, 40126 Bologna, BO, Italy
{stefano.degiorgis2,aldo.gangemi}@unibo.it
[2] ISTC - CNR, Via S. Martino della Battaglia 44, 00185 Roma, RM, Italy
[3] University of Turin, Via Verdi, 8, 10124 Turin, TO, Italy
rossana.damiano@unito.it

Abstract. Values, as intended in ethics, determine the shape and validity of moral and social norms, grounding our everyday individual and community behavior on commonsense knowledge. The attempt to untangle human moral and social value-oriented structure of relations requires investigating both the dimension of subjective human perception of the world, and socio-cultural dynamics and multi-agent social interactions. Formalising latent moral content in human interaction is an appealing perspective that would enable a deeper understanding of both social dynamics and individual cognitive and behavioral dimension. To formalize this broad knowledge area, in the context of ValueNet, a modular ontology representing and operationalising moral and social values, we present two modules aiming at representing two main informal theories in literature: (i) the Basic Human Values theory by Shalom Schwartz and (ii) the Moral Foundations Theory by Graham and Haidt. ValueNet is based on reusable Ontology Design Patterns, is aligned to the DOLCE foundational ontology, and is a component of the Framester factual-linguistic knowledge graph.

Keywords: Moral Values · Knowledge Representation · Frame Semantics · Commonsense Reasoning · Ethics & AI

1 Introduction

Values, as intended in ethics, are part of the "general frame of reference for living" [21], meaning that they are relevant (if not determinant) in our everyday behaviour and decision making, delimiting our conscious self by framing knowledge of what we *should* and what we *desire* [24,27,30,32].

Bilsky and Schwartz investigating the semantics of "values" [3] conceptualize them as similar to social norms, with two important differences: (i) they are not explicitly regulated or formalized, and (ii) their sanction-reward system

S. De Giorgis, A. Gangemi and R. Damiano—Contributed equally.

O. Corcho et al. (Eds.): EKAW 2022, LNAI 13514, pp. 3–18, 2022.
https://doi.org/10.1007/978-3-031-17105-5_1

operates on the emotional layer [28]. In particular they highlight five recurrent features: 1) they are considered as concepts or beliefs; 2) they are related to some desirable states or behaviours; 3) they can be deduced from their realization in specific situations but they transcend them; 4) they are pivotal for selection or evaluation processes and 5) are often organized by relative importance. The last point in particular is commonly shared among studies on the necessary scalar nature of values [37]. Values are furthermore inextricably related to commonsense knowledge and perspectivization, expression of personal positions and freedom of judgement, although the perspective of values differs from deontic reasoning since, in van Fraassen's words [37], *deontology*, or the theory of obligations "deals with what ought to be because it is required by one's station and its duties, by the web of obligations and commitments the past has spun", while, considering social obligations as kantian schemata, product of the human reason and time and space contextually dependent, *axiology*, or the theory of values, "deals with what ought to be because its being so would be good, or at least better, than its alternative". Finally, values are particularly relevant in dynamics of appraisal [34], since our choices and behaviours are typically affected by our values [3,27]; and by the emotions arising from value-driven appraisal dynamics [26]. In social psychology, in fact, the Contempt-Anger-Disgust (CAD) triad model of moral emotions proposed by [28] relates them to specific configurations of values, termed ethics, inspired by Schweder's work [22] on morality from an anthropological perspective. The CAD triad model relates each emotion type to the violation of a specific ethic: Contempt to the Ethics of community, Anger to the Ethics of autonomy, Disgust to the Ethics of Divinity. These ethics can also be seen as a subset of the value-violation dyadic opposition (e.g. Care vs Harm) constituting the Moral Foundations Theory put forth by Haidt and colleagues [19]. Finally, from a neuro-biological perspective [5], "there was a biological blueprinting for the intelligent construction of human values [...] We also believe that a variety of natural modes of biological responses, which include those known as emotions, already embody such values.". This work moves the first steps towards the formalization of the moral and social values as "abstract objects with social capital" [6] and their structure of relations, investigating the domain of subjective human perception as well as socio-cultural dynamics, focusing in particular on models and theories supported by empirical data - namely the Moral Foundation Theory and Basic Human Values - providing for both an ontological axiomatisation, and showing possible inferences. Formalization is inspired by Constructive Descriptions and Situations (CDnS) [9], assuming values as schemas of social norms that enter the complex dynamics of community acceptance and enforcement.

The paper is organized as follows: in Sect. 2 we provide an overview of the resources reused and already existing material, in particular Sect. 2.1 introduces the frame semantics approach adopted to model the whole ValueNet modular ontology, which is described in Sect. 2.2, in particular ValueCore and MFTriggers modules are described in Sects. 2.3 and 2.4. Section 3 explains the new ontological modules introduced in ValueNet in order to formalize the existing theoretical background, in particular Sect. 3.1 is focused on Basic Human Values theory, while Sect. 3.2 is focused on Moral Foundations Theory. Finally, Sect. 4 provides some use-case scenarios for the ontological modules introduced, while Sect. 5 envisions further operationalisation and maintenance of the resource.

This work started being developed originally in the SPICE project context, mentioned in the Acknowledgments, and available on GitHub[1].

2   Preliminaries

In this section, we give an overview of ValueNet[2], then we provide an overview of the theoretical background, in particular the Basic Human Values theory [30] and the Moral Foundations Theory [16], formalised with a frame semantics [10] approach (cf. Sect. 2.2).

2.1   Frame Semantics and Framester

The approach adopted to model ValueNet, and to connect it to the linguistic expression of values, reuses the formal representation of FrameNet frames [8] as formalised [25] in Framester [11]. Frames are defined as cognitive representations of prototypical and recurrent features of events or situations. Lexical units semantically related to some scene are associated with frames, based on their schematic structure. In FrameNet, frames are also explained as *situation types*. In Framester semantics [10] observed/recalled/anticipated/imagined situations are consequently occurrences of frames. For example, representing an apparently simple situation like the moral emotion [36] "feeling ashamed" as a framal structure, it would require some *necessary roles* such as an agent feeling the emotion (experiencer), the emotion itself, and eventually some emotion trigger, but also some optional elements such as the intensity, the physiological manifestation, and some external elements such as the duration of the emotion feeling/state.

Framester provides a formal semantics for frames in a curated linked data version of multiple linguistic resources (e.g. besides FrameNet, WordNet [23], VerbNet [29], etc.); a cognitive layer including MetaNet [12] and ImageSchemaNet [7], connecting conceptual metaphors and image schematic sensorimotor patterns to linguistic resources; factual knowledge bases (e.g. DBpedia [2], YAGO [35], etc.), and ontology schemas (e.g. DOLCE [13]), with formal links between them, resulting in a strongly connected RDF/OWL knowledge graph.

2.2   ValueNet

The ValueNet[3] modular ontology is an extension of Framester, therefore values are modeled as framal structures (also in accordance with CDnS) [9], triggered

---

[1] The SPICE GitHub is available here: https://github.com/spice-h2020/SON.

[2] Some useful prefixes and URIs used in the next sections are available here: https://github.com/StenDoipanni/ValueNet/blob/main/README.md.

[3] ValueNet repository is available here: https://github.com/StenDoipanni/ValueNet.

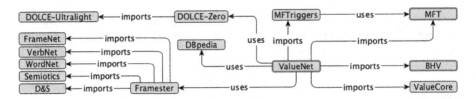

Fig. 1. ValueNet import and usage network.

by other Framester entities, thus enabling a linguistic, cognitive, and factual grounding to values. Its purpose is twofold: (i) it aims at formalizing existing theories about moral and social values, with the goal to create a formal integrated environment, based on the general ValueCore module, described in Sect. 2.3, which allows the integration of theoretical knowledge with experimental data based on a certain theory; (ii) it aims at operationalizing existing theories in order to develop sense-making tools, e.g., a value detector based on MFT, as explained in Sect. 2.4 (Fig. 1).

2.3   ValueCore

The ValueCore module models the notion of "value" as a frame. It reuses the Constructive Description&Situation ontology design pattern [9,14], considering each value of each theory (formalized in separate modules, here we present the Basic Human Values and the Moral Foundations Theory, but Sect. 5 envisions further extensions) as a `fschema:ConceptualFrame`, subclass of `dul:Description`, satisfied by some `vc:ValueSituation`, namely, the realization/occurrence of some prototypical type of event involving some value. Being a core module, it generalises specific notions of value, in order to cover every possible value situation. According to the current literature, the ValueCore module includes three main types of value-driven situations: (i) `vc:ValueAppraisal`: the appraisal of a situation performed by an agent, pivoted by a value; (ii) `vc:ValueCommitment`: the commitment of an agent to a value; and (iii) `vc:ValueRecognition`: the recognition, namely, the plain existence assertion, operated by some agent, of a value in some context. These three types of situation, modeled as framal structures including necessary, optional and external roles, allow to model any type of event including some value, with an increasing detail, proportional to the granularity of the scenario taken in consideration.

The ValueCore module can be explored online[4] or via the Framester endpoint[5].

---

[4] The ValueCore module is available here:
   https://raw.githubusercontent.com/StenDoipanni/ValueNet/main/ValueCore.ttl.
[5] The Framester endpoint is available here: http://etna.istc.cnr.it/framester2/sparql.

2.4 MFTriggers

Another module included in ValueNet is MFTriggers. MFTriggers intends to fill a gap: there is no repository that provides alignments of entities from different semantic layers (lexical units, semantic roles, framal structures, factual entities, etc.), to a social or moral value from any theory. Albeit the Extended Moral Foundations Dictionary [20] has been used to train neural models with the task of detecting moral values, no direct lexical grounding has been provided for any of the elements of the Graham and Haidt's dyadic oppositions, let alone as knowledge graphs.

Therefore, we use MFTriggers to support value detection and value extraction from natural language. MFTriggers introduces a lexical and factual grounding for the Moral Foundations Theory, and therefore for MFT ontological module[6]. Future further operationalisations for other modules and theories (e.g. BHV) are envisioned in Sect. 5.

The automatic values extraction from natural language includes the usage of the FRED tool, a hybrid statistical and rule-based knowledge extraction system to generate RDF and OWL knowledge graphs taking as input directly text from natural language.

The value extraction workflow is composed by three main steps: (i) the first step is to take a sentence and pass it to FRED tool. Figure 2 shows the graph automatically generated for the sentence "*We are organizing a protest against dictatorship.*"[7].

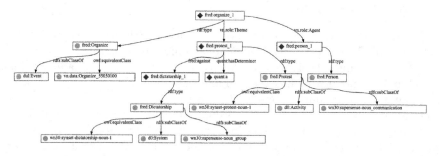

Fig. 2. FRED graph automatically generated for the sentence *We are organizing a protest against dictatorship.*

The second step (ii) consists in taking all the subjects, predicates, and objects of all triples, namely all nodes and arches in the graph, to query the ValueNet graph, in particular, at the current state, the MFTriggers graph, via SPARQL

---

[6] MFTriggers building process is available here: https://github.com/StenDoipanni/ValueNet/tree/main/MFTriggers.

[7] FRED online demo is available at: http://wit.istc.cnr.it/stlab-tools/fred/demo/.

queries, to check if there is any semantic trigger (any entity from FrameNet, WordNet, VerbNet, DBpedia etc.) triggering of some value[8].

Finally, (iii) for each triggering occurrence retrieved, a triple is added to the original graph declaring the activation. In the example above: the WordNet synset `wn:synset-protest-noun-1` triggers `mft:Subversion` and the synset `wn:synset-dictatorship-noun-1` triggers `mft:Oppression`.

3 ValueNet Theoretical Modules

The following sections introduce the BHV and MFT ontological modules, namely the transposition of the Basic Human Values theory and the Moral Foundations Theory in ontological form, showing their main focus and possible inferences. BHV and MFT as theories share some overlaps but start from quite different perspectives, greatly simplified: both theories propose a "universal" model, namely a model which should provide a cultural-agnostic explanation for the whole human value system, and for this reason are modeled in ValueNet. But while MFT adopts a more developmental perspective (explained in detail in Sect. 3.2), BHV considers many socio-behavioral factors. This difference results in both theories having a relational "opposition" of values but while MFT is organized in dyadic oppositions of one value and its violation, BHV circumplex model does not contemplate direct violations, but rather opposition of behavioral focus and attitude.

3.1 Basic Human Values

The Theory of Basic Human Values (BHV) by Shalom Schwartz was proposed as a pan-cultural theory in the 1980s. Its main assumption is that human values are organized in a "value wheel", that is, an ordering structure that organizes values as a circumplex model, dividing them in four quadrants with two opposing axes, and a congruity continuum between adjacent values.

Originally, the model included 10 values [30], but, as shown in Fig. 3, the model was later refined to 19 values in total [32]. BHV relies on the opposition and similarity of values, grouped into macro-categories that are mostly determined by individual personality traits (self-transcendence vs self-enhancement, conservation vs openness to change). This model has inspired the design of a questionnaire (Portrait Values Questionnaire, PTV) which has been employed by a number of studies to explore values across different countries [33]. In recent work [31], Schwartz provides evidence in favour of a pan-cultural arrangement of value priorities.

BHV has been tested on a vast number of subjects across 82 countries. However, one of the main criticism is its top-down approach, establishing the number and taxonomy of values a-priori, and then validating it through dedicated experimentation.

---

[8] Some useful explorative queries are available at: https://github.com/StenDoipanni/ValueNet.

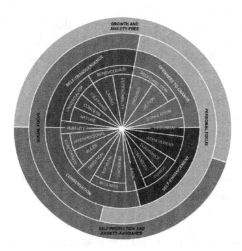

Fig. 3. Basic Human Values circumplex model, image taken from [Giménez, August Corrons, and Lluís Garay Tamajón. "Analysis of the third-order structuring of Shalom Schwartz's theory of basic human values." Heliyon 5.6 (2019): e01797.]

BHV Classes. The ontology takes as source the BHV model reworked as in [15]. It is the attempt to formalize values as an *inner behavioral nudge*, related to outer stimula, towards one (or more) of the four main axes as explained in the following.

The ontology includes 2 top classes representing the "attitude", i.e., a general view of the world, driving some more specific ordering principles; 2 classes representing a "focus", i.e., a taxonomical criterion that addresses the entities (social group, individual, society, class) supposed to profit the most from some value; 4 third order clusters of values, which split the circumplex model in four quadrants, creating diagonal opposition and topical continuity; 12 second order values, namely more specific clusters of values considering a more fine-grained granularity in framing events and situations of the world; and finally 19 first order values, which explicitly state the patient/beneficiary of some value. We list here the ontological classes and axioms, from the most general ones (which in the circumplex model corresponds to most external sectors), to the most specific.

The highest order layer of the circumplex model is formalized as follows:

- `bhv:GrowthAndAnxietyFree`: This is a pro-active attitude, characterizing a self-trascendent view of the world and a higher openness to novelty and change.
- `bhv:SelfProtectionAndAnxietyAvoidance`: this is a more reactive attitude, characterizes as a self-centered view of the world fostering a closer and conservative attitude.

Note that, as shown in Fig. 3, the outer "attitude" ring and the "focus" one have no direct relation between them, being offset from each other, while the main

four quadrants, and, as consequence, the single values, are instead axiomatised
with restrictions on their attitude and focus. Moving therefore one ring inward
into the circumplex model, the "focus" concept is specified in two classes and
modeled as follows:

- bhv:SocialFocus: Focus on social issues and others than self, or focus on
  self, considered as a member of a social community. The focus expresses the
  main beneficiary of the behaviour determined by some Value e.g. the class
  bhv:SelfTrascendence is the superclass grouping all the Values having as
  focus society more than the individual;
- bhv:PersonalFocus: Focus on personal issues and self, both as realization
  of self intended as freedom of thinking and action as well as dominance over
  others.

The third order values layer structures the four main quadrants of the circumplex model. These are modeled as superclasses of more specific value situations,
following Constructive Description and Situation pattern, considering more specific classes of situations as subclasses of more general ones, satisfying more
specific descriptions, subclasses in turn of more general ones. Considering diagonal oppositions (meaning having an opposed value motivation), and according
to their focus and attitude they are:

- bhv:Conservation: This macro category is focused on "preserving stability
  and security", in particular "with the emphasis on subservient self-repression,
  the preservation of traditional practices and protecting stability". In the BHV
  ontological module bhv:Conservation class of value situations is axiomatised
  as:

$$
\begin{aligned}
&\text{SubClassOf:}\\
&((\texttt{attitude some SelfProtectionAndAnxietyAvoidance}) \text{ and}\\
&(\texttt{attitude only SelfProtectionAndAnxietyAvoidance})) \text{ and}\\
&((\texttt{focus some PersonalFocus}) \text{ or } (\texttt{focus some SocialFocus}))
\end{aligned}
\tag{1}
$$

Its opposite quadrant is:

- bhv:OpennessToChange: it consists in readiness for new experience, self centered values which foster physical and intellectual freedom and fulfillment.
  bhv:OpennessToChange class of value situations is axiomatised as:

$$
\begin{aligned}
&\text{SubClassOf:}\\
&((\texttt{attitude some GrowthAndAnxietyFree}) \text{ and}\\
&(\texttt{attitude only GrowthAndAnxietyFree})) \text{ and}\\
&((\texttt{focus some PersonalFocus}) \text{ and}\\
&(\texttt{focus only PersonalFocus}))
\end{aligned}
\tag{2}
$$

The sibling class to
`OpennessToChange` in the circumplex model is:

- `bhv:SelfEnhancement`: it consists in promoting self-interest, often at the expense of others, emphasising the search for personal success and dominance over others. `bhv:SelfEnhancement` class of value situations is axiomatised as:

$$\begin{aligned} \texttt{SubClassOf:} \\ \texttt{((attitude some GrowthAndAnxietyFree) or} \\ \texttt{(attitude some SelfProtectionAndAnxietyAvoidance)) and} \\ \texttt{(focus some PersonalFocus) and (focus only PersonalFocus)} \end{aligned} \quad (3)$$

In the opposed quadrant to `bhv:SelfEnhancement` there is:

- `bhv:SelfTrascendence`: it consists in promoting the well-being of society and nature above one's own interests, highlighting the acceptance of others as equals, as well as a concern for their well-being. `bhv:SelfTrascendence` class of value situations is axiomatised as:

$$\begin{aligned} \texttt{SubClassOf:} \\ \texttt{((attitude some GrowthAndAnxietyFree) and} \\ \texttt{(attitude only GrowthAndAnxietyFree)) and} \\ \texttt{((focus some SocialFocus) and (focus only SocialFocus))} \end{aligned} \quad (4)$$

Finally, the full list of 19 first order BHV values is shown in Fig. 3 and each value class is described in the OWL file[9].

BHV Object Properties. The object properties modeled in BHV module are:

- `bhv:attitude`: this property is used to declare the attitude corresponding to some values, namely `bhv:SelfProtectionAndAnxietyAvoidance` (re-active attitude) vs `bhv:GrothAndAnxietyFree` (pro-active attitude).
- `bhv:focus`: this property is used to declare the focus corresponding to some values, namely `bhv:SocialFocus` vs `bhv:PersonalFocus`.
- `bhv:opposingFocus`: serves the function of modelling oppositions, as described in previous paragraphs and shown in Fig. 3.
- `bhv:opposingValueMotivation`: Following the polarity opposition Conservation vs OpennesToChange and SelfTrascendence vs SelfEnhancement, this property is used to axiomatise all the 4 third order classes of values declaring them as `EquivalentTo`: `opposingValueMotivation some` and `opposingValueMotivation only` the value in the opposite diagonal quadrant.
- `bhv:panCulturallyMoreImportantThat`: to express the eventuality of building a Pan Cultural Baseline For Values Priority.

---

[9] The ontology is available here: https://github.com/spice-h2020/SON/blob/main/SchwartzValues/ontology.owl.

BHV Competency Questions. BHV module allows to answer some CQs according to BHV theory, such as:

1. Is the entity x an instance of some value, according to BHV theory?
2. What values have as focus some `bhv:SocialFocus` or `bhv:PersonalFocus`?
3. What is the `bhv:opposingFocus` of some value?
4. What is the attitude of some value?
5. What is the opposing value motivation for some value?

3.2 Moral Foundations Theory

The Moral Foundation Theory (MFT) is proposed as a cultural-independent theory of moral and social values, inspired by Schweder's et al. work on universal human ethics [22] and tightly related to the investigation of moral emotions, with a particular focus on behavioural neuro-cognitivism. Its agnostic point of view towards cultural dependencies is realized via its dyadic oppositional structure. On one hand, the intension of value-violation dyadic oppositions is supposed to be cultural independent; on the other hand, their extension is dependent on the actual realization of one (or more) dyadic value in some situation of the real world. The model proposed by [17] focuses mainly on single value oppositions, where any pair of opposing values represents the poles of a prescribing/inhibiting dyad. MFT describes six innate moral foundations across cultures and societies:

- Care vs Harm is grounded in the attachment systems and some form of empathy, intended as the ability to not only understand, but also feel, the same feelings as others, thus being able to imagine hypothetical scenarios, in which we are living some positive or negative mental or physical state, which we actually don't live.
- Fairness vs Cheating is grounded in the evolutionary process of reciprocal altruism.
- Loyalty vs Betrayal is grounded in the clans and family-based dimension that for a long time characterized most of our tribal societies. The ability to create links and alliances was a way to increase the surviving percentage possibilities for oneself and his/her close group.
- Authority vs Subversion is grounded in the hierarchical social interactions directly inherited by primates' societies.
- Sanctity vs Degradation is grounded in the CAD triad emotions (Contempt, Anger, Disgust) and the psychology of disgust, it is one of the most spread dyadic oppositions, underlying religious (and not only) notions of living in an elevated, less carnal, more ascetic way. It underlies the idea of "the body as a temple" which can be contaminated by immoral activities and it is foundational for the opposition between soul and flesh.
- Liberty vs Oppression is grounded in feelings and experiences like solidarity, vs. episodes of unjustified violence or liberty restrictions.

Besides its relevance for the investigation of the emotional counterpart of value appraisal and for the cross-cultural investigation of values, MFT has inspired the design of the Moral Foundation Dictionary [18] and, more recently, of the Extended Moral Foundations Dictionary [20], which combine theory-driven elements on moral intuitions with a data-oriented approach. Relationship with the emotion knowledge layer is envisioned as future work in Sect. 5.

Factual situations can evoke some Value, and be opposed by some Violation, creating multi-shaped scenarios, in which the same Event or Action or Entity can evoke different Values and their Violations at the same time.

MFT Classes. The MFT module is light-weighted considering the number of axioms, due to the fact that the whole theory is based on direct dyadic opposition of values and violations. MFT classes are:

- `mft:DyadicOpposition`: this is the superclass for all the value-violation dyads. It `dul:hasComponent` exactly 1 value and exactly 1 violation.
- `mft:Value`: this is the class for "positive" values shaping some behavior, it is subclass of `vc:Value` in the ValueCore module.
- `mft:Violation`: this class represent the violation to some value, they can also be conceived as "negative" values.

MFT Object Properties. The object properties modeled in MFT module are:

- `mft:opposedTo`: some value is opposed to its violation in the dyadic structure. This property is symmetric.
- `mft:violates`: some violation violates some `dul:Norm`.
- `dul:hasComponent`: this property expresses the mereological aspect of some dyad.

MFT Competency Questions. MFT module and (MFTriggers) allow to answer some CQs according to MFT theory, such as:

1. Is the entity x an instance of some value, according to MFT theory?
2. What is the value `mft:opposedTo` some entity x?
3. Is there some value in the sentence y?
4. What is the value profile of (namely the set of values activated by) some word or sentence?

4   Evaluation: BHV and MFT Use Cases

BHV and MFT describe primitive framing of values as descriptions, and are typically associable to real world occurrences (situations), named `vc:ValueSituation`. A value situation presents elements coherent to the conceptualization of BHV or MFT, so that it can answer competency questions mentioned in Sects. 3.1 and 3.2.

To allow an evaluation of the ontological module we propose here a scenario answering CQs mentioned in Sect. 3.1 and 3.2, involving at the same time three types of value situations according to ValueCore module, namely `vc:ValueRecognition`, `vc:ValueAppraisal` and `vc:ValueCommitment`. Furthermore, the methodology mentioned in Sect. 2.4 is extensively tested in [1] where a graph-based Value Detector is compared to (and equals the performance of) state of the art Zero-shot learning method for a Value Detection task.

Value Scenario. UserA and UserB are visiting an art gallery and see a painting depicting Pietro Micca ("Pietro Micca nel punto di dare fuoco alla mina volge a Dio e alla Patria I suoi ultimi pensieri" - "Pietro Micca, the moment before setting fire to the bomb, directs his thoughts to God and his motherland") by Andrea Gastaldi. Pietro Micca is described as an Italian patriot who gave his life to save the to-be-born state of Italy, igniting some dynamite to detonate a tunnel that was being invaded by enemy soldiers.

Pietro Micca's action can be modeled as a `vc:ValueCommitment` situation, nested in two different interpretations of UserA and UserB which can be modeled as `vc:ValueRecognition` situations, and for each of them would be possible to express the appraisal and the desirability of some action for both Users in a `vc:ValueAppraisal` situation[10].

4.1   BHV Inferences

UserA declares to be proud of the action made by Pietro Micca, sharing with him the value "Patriotism". UserB disagrees considering more important "Self Preservation" than sacrificing one's own life to defend the country. Thanks to BHV module and the lexical tokens linked to the first order values, "Patriotism" is inferred as being an instance of both `bhv:Societal` and `bhv:Caring` (see Sect. 3.1 CQ1), subclass of `bhv:Security` and `bhv:Benevolence` and therefore having as opposing value motivations (namely being in the quadrant opposed to) both `bhv:SelfEnhancement` and `bhv:OpennessToChange` (see Sect. 3.1 CQ5), while "Self-Preservation" is an instance of `bhv:Action`, subclass of `bhv:SelfDirection`.
We can infer that UserA's instance of "Patriotism" has `bhv:focus` some `bhv:SocialFocus` (see Sect. 3.1 CQ2) and attitude both `bhv:SelfProtectionAndAnxietyAvoidance` and `bhv:GrowthAndAnxietyFree` (see Sect. 3.1 CQ4); while for UserB's value instance we can infer that it has some `bhv:PersonalFocus`, opposed to UserA's focus (see Sect. 3.1 CQ3) and `bhv:GrowthAndAnxietyFree` attitude.

Similar scenarios to the one proposed here in natural language are available serialized in turtle syntax both on the ValueNet and SPICE project GitHub. Finally, a knowledge graph of semantic triggers operationalizing BHV theory in

---

[10] We do not provide details about the ValueCore possible inferences here since it's not the main focus, but further details are available on the ValueNet GitHub:
https://github.com/StenDoipanni/ValueNet
and on the SPICE project GitHub:
https://github.com/spice-h2020/SON/blob/main/SchwartzValues/Schwartz_scenario.ttl.

order to provide an automatic extraction of value situations and value detection from natural language (as for MFT, in MFTriggers graph described in Sect. 4.2) is being developed and is mentioned as future work in Sect. 5.

4.2   MFT Inferences

UserA declares to be proud of the Action made by Pietro Micca, focusing on the result of this action, namely the Liberty of Italy. UserB disagrees, considering more important Pietro Micca's life than any victory in war, in fact she/he considers it useless to sacrifice oneself for any country. Thanks to MFTriggers "LibertyOfItaly" is inferred as triggering a `mft:Liberty` value Situation and "PietroMiccaSacrifice" is inferred as triggering an `mft:Harm` situation (see Sect. 3.2 CQ3-CQ4). Thanks to the MFT dyadic model, "LibertyOfItaly" is inferred as being an instance of `mft:Liberty` (see Sect. 3.1 CQ1), while "CareOfPietroMicca" is an instance of `mft:Care`, being opposed to "PietroMiccaSacrifice", which is an instance of `mft:Harm` (see Sect. 3.1 CQ2).

5   Conclusions and Future Improvement

We presented here the BHV and MFT theoretical modules integrated in the ValueNet ontology. The ontology is an ongoing attempt to formalize different perspectivizations depending on the cognitive framing that agents make in relation to some stimulus. The current version includes the ValueCore and MFTriggers modules, as well as the newly introduced and presented here BHV and MFT ontologies.

Future developments on the theoretical side include the introduction of new theoretical modules, such as Curry's "Moral Molecules" [4] theory[11]; while on the operational side they include the semantic triggers knowledge graphs generation, starting from resources like Schwartz's Portrait Value Questionnaire [31]. Furthermore, another interesting direction of research would be to conjugate this symbolic approach with BERT-like pre-trained models. On a parallel research direction ontological modules formalizing theoretical Emotion theories and generating semantic triggers knowledge graphs are being developed and introduced as an Emotion knowledge layer in the Framester resource. Future developments will be to conjugate these two intertwined layers in more complex formal semantics representations.

Acknowledgements. This work is funded by the SPICE EU H2020 Project 870811 within the program: SOCIETAL CHALLENGES - Europe In A Changing World - Inclusive, Innovative And Reflective Societies.

---

[11] Curry's ontological module is available at:
https://github.com/StenDoipanni/ValueNet/tree/main/MoralMolecules.

References

1. Asprino, L., Bulla, L., De Giorgis, S., Gangemi, A., Marinucci, L., Mongiovì, M.: Uncovering values: detecting latent moral content from natural language with explainable and non-trained methods. In: Proceedings of ACL 2022 Workshop DEELio on Knowledge Extraction and Integration for Deep Learning Architectures (2022)
2. Auer, S., Bizer, C., Kobilarov, G., Lehmann, J., Cyganiak, R., Ives, Z.: DBpedia: a nucleus for a web of open data. In: Aberer, K., et al. (eds.) ASWC/ISWC -2007. LNCS, vol. 4825, pp. 722–735. Springer, Heidelberg (2007). https://doi.org/10.1007/978-3-540-76298-0_52
3. Bilsky, W., Schwartz, S.H.: Values and personality. Eur. J. Pers. **8**(3), 163–181 (1994)
4. Curry, O.S., Alfano, M., Brandt, M.J., Pelican, C.: Moral molecules: morality as a combinatorial system. Rev. Phil. Psych., 1–20 (2021). https://doi.org/10.1007/s13164-021-00540-x
5. Damasio, A.: The neurobiological grounding of human values. In: Changeux, J.-P., Damasio, A.R., Singer, W., Christen, Y. (eds.) Neurobiology of Human Values, pp. 47–56. Springer, Heidelberg (2005). https://doi.org/10.1007/3-540-29803-7_5
6. De Giorgis, S., Gangemi, A.: Exuviae: an ontology for conceptual epistemic comparison. In: 2022 Proceedings of the 6th International Conference on Graphs and Networks in the Humanities, Amsterdam, Netherlands (2022, accepted)
7. De Giorgis, S., Gangemi, A., Gromann, D.: ImageSchemaNet: formalizing embodied commonsense knowledge providing an image-schematic layer to Framester. Semant. Web J. (2022, forthcoming)
8. Fillmore, C.J.: Frames and the semantics of understanding. Quaderni di semantica **6**(2), 222–254 (1985)
9. Gangemi, A.: Norms and plans as unification criteria for social collectives. Auton. Agent. Multi-Agent Syst. **17**(1), 70–112 (2008)
10. Gangemi, A.: Closing the loop between knowledge patterns in cognition and the semantic web. Semant. Web **11**(1), 139–151 (2020)
11. Gangemi, A., Alam, M., Asprino, L., Presutti, V., Recupero, D.R.: Framester: a wide coverage linguistic linked data hub. In: Blomqvist, E., Ciancarini, P., Poggi, F., Vitali, F. (eds.) EKAW 2016. LNCS (LNAI), vol. 10024, pp. 239–254. Springer, Cham (2016). https://doi.org/10.1007/978-3-319-49004-5_16
12. Gangemi, A., Alam, M., Presutti, V.: Amnestic forgery: an ontology of conceptual metaphors. In: Proceedings of the 10th International Conference, FOIS 2018, pp. 159–172 (2018). https://doi.org/10.3233/978-1-61499-910-2-159
13. Gangemi, A., Guarino, N., Masolo, C., Oltramari, A.: Sweetening WORDNET with DOLCE. AI Mag. **24**(3), 13 (2003)
14. Gangemi, A., Mika, P.: Understanding the semantic web through descriptions and situations. In: Meersman, R., Tari, Z., Schmidt, D.C. (eds.) OTM 2003. LNCS, vol. 2888, pp. 689–706. Springer, Heidelberg (2003). https://doi.org/10.1007/978-3-540-39964-3_44
15. Giménez, A.C., Tamajón, L.G.: Analysis of the third-order structuring of shalom Schwartz's theory of basic human values. Heliyon **5**(6), e01797 (2019)
16. Graham, J., et al.: Moral foundations theory: the pragmatic validity of moral pluralism. In: Advances in Experimental Social Psychology, vol. 47, pp. 55–130. Elsevier (2013)

17. Graham, J., Haidt, J., Nosek, B.A.: Liberals and conservatives rely on different sets of moral foundations. J. Pers. Soc. Psychol. **96**(5), 1029 (2009)
18. Graham, J., Nosek, B.A., Haidt, J., Iyer, R., Koleva, S., Ditto, P.H.: Mapping the moral domain. J. Pers. Soc. Psychol. **101**(2), 366 (2011)
19. Haidt, J.: The Righteous Mind: Why Good People are Divided by Politics and Religion. Vintage (2012)
20. Hopp, F.R., Fisher, J.T., Cornell, D., Huskey, R., Weber, R.: The extended moral foundations dictionary (eMFD): development and applications of a crowd-sourced approach to extracting moral intuitions from text. Behav. Res. Meth. **53**(1), 232–246 (2021)
21. Maddi, S.R.: Personality Theories: A Comparative Analysis. Thomson Brooks/Cole Publishing Co. (1996)
22. Mahapatra, M., Park, L.: The "big three" of morality (autonomy, community, divinity) and the "big three" explanations of suffering, p. 119 (1997)
23. Miller, G.A.: WordNet: An Electronic Lexical Database. MIT Press (1998)
24. Nowack, D., Schoderer, S.: The role of values for social cohesion: theoretical explication and empirical exploration. German Development Institute, Discussion Paper 6 (2020)
25. Nuzzolese, A.G., Gangemi, A., Presutti, V.: Gathering lexical linked data and knowledge patterns from FrameNet. In: Proceedings of the 6th International Conference on Knowledge Capture, pp. 41–48. ACM (2011)
26. Prinz, J.J., Nichols, S.: Moral emotions. In: The Moral Psychology Handbook (2010)
27. Rokeach, M.: The Nature of Human Values. Free Press (1973)
28. Rozin, P., Lowery, L., Imada, S., Haidt, J.: The cad triad hypothesis: a mapping between three moral emotions (contempt, anger, disgust) and three moral codes (community, autonomy, divinity). J. Pers. Soc. Psychol. **76**(4), 574 (1999)
29. Schuler, K.K.: VerbNet: a broad-coverage, comprehensive verb lexicon. University of Pennsylvania (2005)
30. Schwartz, S.H.: Universals in the content and structure of values: theoretical advances and empirical tests in 20 countries. In: Advances in Experimental Social Psychology, vol. 25, pp. 1–65. Elsevier (1992)
31. Schwartz, S.H.: An overview of the Schwartz theory of basic values. Online Readings Psychol. Cult. **2**(1) (2012). ISSN 2307-0919
32. Schwartz, S.H., et al.: Refining the theory of basic individual values. J. Pers. Soc. Psychol. **103**(4), 663 (2012)
33. Schwartz, S.H., Melech, G., Lehmann, A., Burgess, S., Harris, M., Owens, V.: Extending the cross-cultural validity of the theory of basic human values with a different method of measurement. J. Cross Cult. Psychol. **32**(5), 519–542 (2001)
34. Strohminger, N., Kumar, V.: The Moral Psychology of Disgust. Rowman & Littlefield (2018)
35. Suchanek, F.M., Kasneci, G., Weikum, G.: YAGO: a core of semantic knowledge. In: Proceedings of the 16th International Conference on World Wide Web, pp. 697–706 (2007)
36. Tangney, J.P., Stuewig, J., Mashek, D.J.: Moral emotions and moral behavior. Ann. Rev. Psychol. **58**, 345 (2007)
37. Van Fraassen, B.C.: Values and the heart's command. J. Philos. **70**(1), 5–19 (1973)

Extending Ontology Engineering Practices to Facilitate Application Development

Paola Espinoza-Arias(✉) , Daniel Garijo , and Oscar Corcho

Ontology Engineering Group, Universidad Politécnica de Madrid, Madrid, Spain
{pespinoza,ocorcho}@fi.upm.es, daniel.garijo@upm.es

Abstract. Ontologies define data organization and meaning in Knowledge Graphs (KGs). However, ontologies have generally not been taken into account when designing and generating Application Programming Interfaces (APIs) to allow developers to consume KG data in a developer-friendly way. To fill this gap, this work proposes a method for API generation based on the artefacts generated during the ontology development process. This method is described as part of a new phase, called ontology exploitation, that may be included in the last stages of the traditional ontology development methodologies. Moreover, to support some of the tasks of the proposed method, we developed OATAPI, a tool that generates APIs from two ontology artefacts: the competency questions and the ontology serialization. The conclusions of this work reflect that the limitations found in the state-of-the-art have been addressed both at the methodological and tooling levels for the generation of APIs based on ontology artefacts. Finally, the lines of future work present several challenges that need to be addressed so that the potential of KGs and ontologies can be more easily exploited by application developers.

Keywords: Ontology Engineering · Application Development · Application Programming Interface · Ontology Artefacts

1 Introduction

Over recent years, Knowledge Graphs (KGs) have been generated and adopted by many organizations to integrate data, facilitate interoperability, and generate new insights and recommendations. KGs are commonly structured according to ontologies, which allow data to be unambiguously defined with a shared and agreed meaning, as well as to infer new knowledge. However, despite their adoption, KGs are still challenging to consume by application developers.

On the one hand, developers face a *production-consumption challenge*: there is a gap between the ontology engineers who design an ontology and may intervene in KG creation and the application developers who want to consume its contents [7]. Ontologies may be complex, and the resources generated during their development (use cases, requirements, etc.) are often not made available to their users (e.g. application developers). As a result, developers usually need to duplicate some

O. Corcho et al. (Eds.): EKAW 2022, LNAI 13514, pp. 19–35, 2022.
https://doi.org/10.1007/978-3-031-17105-5_2

of the effort already done by ontology engineers when they were understanding the domain, interacting with domain experts, taking modeling decisions, etc. On the other hand, application developers face *technical challenges*: many of them are not familiar with Semantic Web standards such as OWL and SPARQL, and hence those KGs that are exclusively based on Semantic Web technologies remain hardly accessible to them [18]. Developers (and in particular application developers) are mostly used to data representation formats like JSON and Application Programming Interfaces (APIs) for accessing data.

In order to address both production-consumption and technical challenges, multiple approaches have been proposed by the Semantic Web community, ranging from Semantic RESTful APIs [15] which are compatible with Semantic Web and REST; to tools to create Web APIs on top of SPARQL endpoints [1,3,4,12]. Outside the Semantic Web community, approaches like GraphQL[1] are gaining traction among developers due to their flexibility to query and retrieve data from public endpoints. However, generating APIs based on ontology artefacts has received less attention so far. These artefacts are any intermediate or final resources generated during the ontology development process (e.g. competency questions, SPARQL queries, ontology serialization, etc.).

The main goal of this work is to ease KG data consumption by application developers who are not experts in ontologies, while reusing some of those intermediate or final resources of the ontology development process. Therefore, we focus on the following research questions: RQ1: *Is it possible to generate APIs based on the artefacts created by ontology engineers during the ontology development process?*, RQ2: *Is it possible to automatically generate APIs that are indistinguishable from APIs developed by application developers in real use cases?* To answer our RQs, we propose a novel **method for API generation based on ontology artefacts**, which proposes a set of activities to define, specify, implement, validate, and deploy APIs. In addition, we develop a **proof-of-concept tool for supporting API generation**, which allows building a set of APIs and SPARQL queries based on two ontology artefacts: competency questions and ontology serialization.

The remainder of this paper is organized as follows. We begin by describing the related work (Sect. 2). Then, we present the proposed method in Sect. 3 and how to support its execution in Sect. 4. In Sect. 5, we present the results of our evaluation. Finally, we discuss future work and conclusions in Sect. 6.

2   Related Work

We consider in the related work the most relevant and well-known methodologies for ontology development with a special focus on the stages and activities they propose, and the ontology artefacts they produce. Several approaches have been proposed for developing ontologies (relevant surveys are reported in [2,10,11]). There are heavyweight methodologies that require time and resource-consuming activities (e.g. METHONTOLOGY [6], On-To-Knowledge [16], or NeOn [17]), and

---

[1] https://spec.graphql.org/June2018.

lightweight methodologies based on agile techniques that allow building ontologies that are always ready to be used (e.g. SAMOD [13] or LOT [14]). These methodologies propose, at a lower or higher level of detail, similar core phases to develop ontologies. At the beginning of the development, the methodologies propose identifying the requirements, expressed as competency questions (CQs), that the ontology must meet. Next, they propose generating a model or intermediate representation of the ontology containing the terms and the relationships between them. Then, they propose formalizing the model (using an ontology implementation language such as OWL) and, finally, testing whether the formal representation of the ontology answers the CQs. Also, best practices for ontology publishing[2] suggest publishing the formal model along with its human-readable documentation. At the end, some methodologies identify a maintenance stage to allow fixing bugs and updating the ontology. In addition, several ontology artefacts are generated during the development process suggested by these methodologies. For example, the competency questions (CQs) specify ontology requirements, the ontology serialization formalizes the ontology, the glossary of terms extracts key terms and definitions from documents or data, etc. Ontology development methodologies have evolved to develop these artifacts through documented phases, activities, tasks, guidelines, and techniques for ontology development. However, there is a lack of detail on how to use an ontology after it has been generated (except for usage examples in its documentation).

Moreover, as part of the related work, we conducted a study, reported in [5], in which we analyzed existing approaches, techniques, and tools for the generation of APIs from ontologies. Our findings revealed that most of the tools and technologies do not consider ontologies or ontology artefacts for building APIs. In fact, most of the approaches we analyzed generate APIs from SPARQL queries. Only OBA [8] and OWL2OAS[3] allow generating an OAS[4] document from the OWL ontology. Moreover, OBA automatically generates SPARQL query templates needed to execute CRUD operations, and it also provides the server-side functionality for the API. Furthermore, we found that in all the tools and technologies analyzed the effort was focused on the technological support to automatically generate APIs rather than on a methodology to design them. Thus, we can state that there is not a methodological approach to build APIs taking as input the ontology artefacts produced during the ontology development process.

3   Method for API Generation Based on Ontology Artefacts

Existing ontology development methodologies (e.g. LOT, NeOn, among others mentioned in Sect. 2) usually involve phases such as requirements specification, ontology implementation, ontology publication, and maintenance. However, none of these methodologies pays much attention to how the ontology will be consumed

---

[2] https://www.w3.org/TR/swbp-vocab-pub.

[3] https://github.com/RealEstateCore/OWL2OAS.

[4] OpenAPI Specification (OAS) https://swagger.io/specification.

after it has been generated. To fill this gap, we propose a new *ontology exploitation* phase encompassing any task where the ontology must be used; therefore, it may include tasks such as RDB2RDF mapping definition, RDF data generation, data consumption mechanisms provision, among others.

In this work, we focus on providing data consumption mechanisms through APIs, as none of the well-established methodologies provide details on how to generate APIs for KG data consumption. Thus, the ontology exploitation phase describes the *API generation* process. Figure 1 illustrates this phase as an extension of the LOT methodology.[5] However, this phase can be adopted by any methodology as it makes use of artefacts that are commonly generated in all these methodologies and it should be considered at the end of the development process. As shown in Fig. 1, the actors involved in the ontology exploitation phase are ontology developers (aka ontology engineers) and application developers. In addition, the inputs of the phase are the ontology artefacts produced during the previous phases of the ontology development process, and the output is an API.

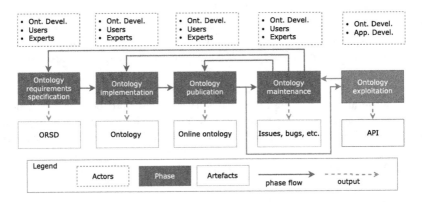

Fig. 1. Ontology exploitation phase as an extension of the LOT methodology. New elements are illustrated in magenta.

To define an API, our method proposes a set of activities to build APIs from ontology artefacts. This method is inspired by the workflow for designing and building Web APIs presented in [9]. Figure 2 summarizes the five activities of our method. Below we describe each activity and to illustrate its execution we present some examples using two artefacts of an ontology for the representation of the local businesses of a municipality developed in the context of the Open Cities (https://ciudades-abiertas.es) project. The artefacts from the Local Business Census ontology are: its ontology serialization[6] and its CQs.[7]

---

[5] For readability, we call the LOT's activities phases. We describe the methodology as a set of phases involving activities, these activities involve several tasks, and so on.

[6] http://vocab.ciudadesabiertas.es/def/comercio/tejido-comercial.

[7] https://github.com/CiudadesAbiertas/vocab-comercio-censo-locales/tree/master/requirements.

*1) **API Design:*** This activity focuses on deciding how the API will behave and defining the resources along with the operations that the API will provide. To this end, we propose a set of tasks to guide ontology engineers and application developers in making these decisions and defining resources and operations taking ontology artefacts as the main input. Figure 3 shows the tasks of this activity and their inputs and outputs. Each task is described below.

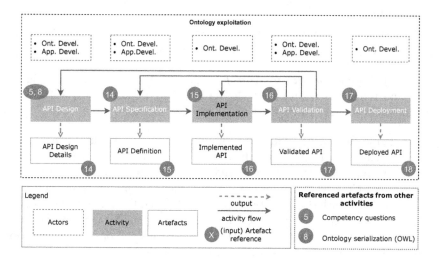

Fig. 2. Activities of the method for API generation based on ontology artefacts. This figure follows the convention defined by the LOT methodology.

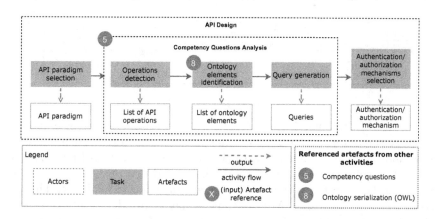

Fig. 3. Tasks and subtasks involved in the API design activity

*1.1) **API Paradigm Selection:*** This task aims to choose what style the API will follow. The selected paradigm may be a request-response one such as those oriented to a resource-style, hypermedia-style, query-style, among others. The selection on which paradigm to use depends on the decision of ontology engineers and application developers, after an analysis of the existing paradigms and how they fit in the requirements that need to be addressed by the ontology, maintainability, a specific application behaviour, etc. In the Semantic Web community the most common paradigms adopted are REST (resource-style) and GraphQL (query-style), as we have detected in our study reported in [5]. Thus, this method describes how to carry out the activities and tasks according to both paradigms.

Example

In this first task, let us assume we select the resource-oriented REST paradigm to generate the API of the Local Business Census ontology.

1.2) Competency Questions Analysis: This task focuses on analyzing the CQs to find the terms required for the API design. To this end, the following three substasks should be executed:

- **Operations detection.** This subtask aims to identify which operations will be implemented in the API according to the intent of the CQs. To detect operations, the ontology engineer must analyze specific terms that request something in the CQs. Table 1 shows some examples of common terms and their correspondence with the operations of the REST and GraphQL paradigms. In general, CQs consider read-only operations rather than data change requests. However, if a CQ intends to perform data changes the operation detection should also consider terms related to write operations.

Table 1. Common competency questions terms denoting operations in APIs.

Paradigm Term	REST	GraphQL
get, list, which, what, who, where, when, obtain, give	GET	query
add, insert	POST	mutation
update, change	PUT	mutation
delete, remove	DELETE	mutation

In the case of CQs intended to solve boolean or counting questions a different strategy is required. Table 2 shows examples of common terms defining these CQs and their correspondence with the operations of each API paradigm. In general, to solve these queries, CQs can be related to a read operation whose resulting data must be processed to deliver the expected answers. Thus, to solve counting questions, API consumers may count the resulting data on the client-side to get the number of elements. As for boolean questions, API consumers may analyze the resulting data to determine whether the query response is true (more than zero results), or false (zero results).

Table 2. Example of operations detection for counting and boolean queries

Term	Query type	REST	GraphQL
how many	counting	GET	query
is, was, were	boolean	GET	query

Example

For the CQs analysis, we choose one question from the set of CQs of the Local Business Census ontology. Then, we analyze each term of the CQ and detect a coincidence with one of the terms presented in Table 1. The following image shows the CQ, the term denoting an operation (Give), and the detected operation (GET):

GET
↑
Give me the details of the terrace that belongs to a local business.

- **Ontology elements identification**. This subtask aims to distinguish which ontology elements are required by the API. To do so, it uses the ontology serialization as input to verify whether the labels of the ontology elements match the terms identified in the CQs. This verification is essential to detect how these elements are related and, depending on the operation, the structure of the expected API input/output. An alternative way to identify the ontology elements may be to analyze the fragment identifier of their URIs. This alternative can be applied when URIs are defined with short and compact names in natural language. However, this alternative may not be applicable when URIs follow an opaque strategy (which obfuscates the name of the ontology elements) because it hinders the elements identification.

Moreover, in those cases where the terms identified do not match ontology elements it would be necessary to check for their synonyms. To do this, the ontology engineer may reuse the glossary of terms generated at the beginning of the ontology development. Thus, some terms found in the CQ can be compared with those in the glossary to find the corresponding ontology elements. Another option for identifying terms would be to use existing open glossaries (domain dependent or independent).

Finally, when the selected API paradigm is REST it will be necessary to use the elements identified in this subtask to define the API paths required for each CQ. To this end, we designed a set of rules for naming REST APIs based on ontology artefacts. These rules are described in our GitHub repository.[8]

---

[8] https://github.com/oeg-upm/oatapi/blob/main/Additional%20Resources.

Example

To identify the ontology elements, we analyze the CQ and the ontology serialization. In the figure below, we illustrate the ontology serialization as an excerpt of the diagram of the Local Businesses Census ontology.

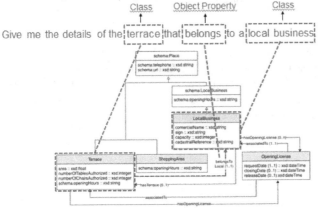

As shown in this figure, some terms from the CQ match the labels of the ontology serialization, and, as a result, we identified the `Terrace` and `LocalBusiness` classes and the `belongsToLocal` object property. Furthermore, as we are defining a REST API we generate the API path required for the CQ. Taking the resulting ontology elements, we applied the rules for naming REST APIs and we built the following path: `/local-businesses/{local-business-id}/terraces`.

– **Query generation.** This subtask aims to define the query to manage the data and the functionality required by the CQ. To this end, the SPARQL queries generated for the ontology validation may be reused, if available. This validation is common in ontology development methodologies as it allows ontology engineers to verify whether the ontology meets the requirements against a proof-of-concept dataset. But if queries had not been defined or are not available they must be generated. Thus, for query generation, two steps should be performed. First, take the operation, detected in the first subtask, and determine what SPARQL operation should be included in the query. Second, match the ontology elements, detected in the second subtask, with the query structure and format the query accordingly. The query structure should be defined according to the CQ intention and its expected answer.

> **Example**
>
> We manually build the SPARQL query to fetch the data, as follows:
>
> ```
> 1  PREFIX escom:<http://vocab.ciudadesabiertas.es/def/comercio/
>        tejido-comercial#>
> 2  CONSTRUCT { ?terrace ?predicate ?object }
> 3    WHERE { {
> 4      SELECT DISTINCT ?terrace
> 5        WHERE {?terrace escom:perteneceALocal ?local-business-id}}
> 6    ?terrace ?predicate ?object }
> ```
>
> Note that the ?local-business-id variable will be replaced by the value specified in the request as it corresponds to the {local-business -id} template defined in the API path. Also, as the operation is GET, the SPARQL operation corresponds to SELECT. This query includes a CONSTRUCT statement, since the result of the request will be composed of all the values of the properties of each instance found.

1.3) Authentication and Authorization Mechanisms Selection: This task aims to define which security mechanisms the API must handle. Depending on operations and data that the API will manage, it may be necessary to provide a security level to control who is allowed and what functions they may execute. For example, it may be necessary to restrict personal data access, or write operations may be allowed only to certain users. The security methods to be provided may be basic (e.g. HTTP requests including username and password) or more reliable (e.g. OAUth 2.0 protocol which is based on the provision of a token access).

> **Example**
>
> Finally, since this API does not involve confidential data and the CQs correspond to read operations, we decided not to use a security mechanism.

2) API Specification: This activity intends to describe the decisions made in the API design. To do so, a common practice is to use an Interface Description Language (IDL) that allows documenting various aspects of the API according to a syntax provided in a machine-readable format (e.g. OAS). Using an IDL allows to quickly generate API documentation because its syntax can be translated, for example, into an HTML document that may include interactions showing how the API works. However, other specifications[9] can be used as long as they provide details about exposed resources, allowed methods, result formats, expected inputs/outputs, among others. It is important to provide a good specification to help application developers learn and understand how the API works.

---

[9] An API specification template we have developed is available in https://github.com/oeg-upm/oatapi/tree/main/Additional%20Resources.

Example

Assume that we write the API details according to the specification template we designed. The following table presents an extract of this template filled in with the details of the API path in terms of the allowed operation, its input, and the expected result. The full specification of this example is available in our repository along with our API Specification template.

API path	Operation	Input	Output
/local-businesses/ {local-business-id}/ terraces	GET	local-business-id	an array of terrace instances (each instance will contain its attribute values of area, table number, opening hours, number of authorized chairs, among others.)

---

3) API Implementation: This activity refers to building the API functionality according to its design and specification. Thus, it is necessary to build the server-side logic to handle each request and deliver the response from the SPARQL endpoint. Also, if a security mechanism is required, the API behavior must be configured based on the user executing the request. Figure 4 illustrates the interactions that can occur between the client and the endpoint. In this figure, the API acts as a middle layer that allows clients to make requests. These requests are processed by the server to query the endpoint, which returns the data. Last, the server delivers the data to the client in the appropriate format via the API.

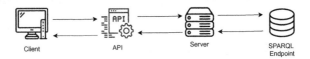

Client   API   Server   SPARQL Endpoint

Fig. 4. Interactions between a client and SPARQL endpoint

Example

Suppose we develop the server-side functionality to handle the API calls and provide the expected responses. This functionality includes executing SPARQL queries according to each API call, serializing the results to deliver the response in a developer-friendly format (e.g. JSON), and configuring the messages to retrieve when executing the requests.

---

4) API Validation: This activity is carried out jointly by the ontology engineer and application developer, and it is concerned with testing whether the API calls perform the requested behaviour, which is determined by the answers of the CQs. If this validation fails it will be necessary to go back to the previous activities to review and refine the design, specification or implementation of the API.

Example

Suppose the server we implemented in the previous activity is up, and we develop a test script that acts as a client to make API calls against this server. The following figure illustrates the API validation through the interactions to be performed between the client, server and SPARQL endpoint.

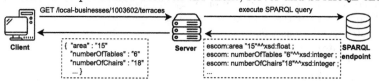

First, the client executes the GET call requesting data from the local business 10003602. Then, the server receives this call and executes the SPARQL query that was defined earlier in our example. Next, the SPARQL endpoint returns the attributes and values of the terraces belonging to this local business (for simplicity, few attributes and values are shown in Turtle). Lastly, the server fetches data from the endpoint, formats that data into the requested serialization (JSON), and delivers it to the client. This validation is successful as the response contains the expected data described in the output column of our API specification example.

5) API Deployment: This last activity refers to making the API available online. Moreover, in this activity it is important to provide access to the API documentation to help developers learn and understand how the API works. Thus, the API specification, represented in a human-readable format, should also be made available. Finally, it is advisable to make the API findable on the Web to ease its promotion and discovery. To this end, ontology engineers may publish the API in registries (e.g., Programmable Web[10]).

Example

Finally, suppose we put the server on-line, make the API specification available on the Web, and register the API in the Programmable Web registry.

4 Supporting the API Generation Process

To automate some activities and tasks of our method, we developed OATAPI (Ontology Artefacts to API). This proof-of-concept tool[11] takes as input a set of CQs and the ontology serialization. Then, it parses both artefacts and delivers a set of REST API paths and SPARQL queries that allow getting data to solve the CQs. We decided to build REST APIs since REST allows us to directly use SPARQL queries to access data that is hosted on traditional SPARQL endpoints. Thus, OATAPI performs by default the paradigm selection from the API design

---

[10] https://www.programmableweb.com.

[11] OATAPI is publicly available in https://github.com/oeg-upm/oatapi.

activity. This tool also automates the CQ analysis task from the API design, which is formalized in Algorithm 1.

Algorithm 1 : Pseudocode for the competency questions analysis
---
Input: Competency questions (CQs) and ontology serialization (onto)
Output: API paths (APIpaths) and SPARQL queries (queries)
 1: loadCompetencyQuestions(CQs);
 2: loadOntology(onto);
 3: **for all** cq in CQs **do**
 4:    $CQ \longleftarrow cqBreakdown(cq)$;
 5:    $operation \longleftarrow operationIdentification(CQ)$;
 6:    $ontologyElements \longleftarrow ontologyElementsIdentification(onto)$;
 7:    $APIpaths, nodesFound \longleftarrow APIpathGeneration(ontologyElements)$;
 8:    $queries \longleftarrow queryGeneration(operation, nodesFound)$;
 9: **end for**
10: return APIpaths, queries;

This algorithm begins by loading the CQs and ontology serialization into OATAPI. Then, it executes the following steps:

1. **Competency questions breakdown.** It consists of splitting each CQ into pieces to be able to analyze its terms, and tagging each piece according to its respective part-of-speech and dependency tree labels. Each piece considers only the base form of the terms to perform the analysis.
2. **Operations identification.** It detects which pieces from step 1 match terms denoting operations. To that end, the algorithm compares each piece with the terms presented in Table 1.
3. **Ontology elements identification.** It identifies the elements in two steps:
   (a) Load the ontology elements into a directed multigraph. Classes are considered as nodes and object properties as edges. Thus, OATAPI requires that object properties contain domain and range restrictions such that edges contain information about the node they come from and go to.
   (b) Take each CQ's piece and check if it matches the label of the ontology elements. If the ontology does not contain labels, the algorithm checks the fragment URI identifier of the ontology elements (only in case the fragment is defined with names in natural language).
4. **API path generation.** It builds the paths by executing the following stages:
   (a) Search for coincidences between the ontology elements identified in step 3 and the nodes from the multigraph.
   (b) Obtain the shortest path between the nodes found in step 4(a).
   (c) Take the nodes of the shortest path and generate the API path. To build the path, the algorithm considers the order of the nodes and follows the rules for naming REST APIs based on ontology artefacts we designed.
5. **Query generation.** It builds SPARQL CONSTRUCT queries using the nodes obtained in step 4(b) and the operations identified in step 2.

It is worth mentioning that the stages described in step 4 present a common case when there are matches between ontology elements and nodes. However, there are cases where only one node is detected, where an edge matches an object property, or both. Moreover, step 5 was also described for a common case when a shortest path is found; but, the query structure will change when other cases occur (e.g. when only one or two nodes are found). These cases are supported by OATAPI, but have omitted from the manuscript due to space constraints.

5   Evaluation

We evaluated[12] our work with two experiments. The first one describes a survey asking users about the paths generated by our approach, while the second one compares our paths against a baseline built by hand.

5.1   Experiment 1

Experiment Settings. For this evaluation, we generated a corpus containing: 1) a set of CQs and its ontologies (from different domains), and 2) pre-existing API paths generated by developers in real use cases for these CQs and ontologies. In total, we gathered 3 ontologies, 9 CQs, and 9 API paths. Next, taking the CQs and ontologies from our corpus as input, we ran OATAPI to automatically build their API paths, and also included these resulting paths in our corpus. Then, to evaluate the API paths gathered in our corpus, we created a questionnaire that we used to ask participants about: a) their background, b) their skills on running basic GET operations in REST API calls, and c) for each CQ, whether the API paths presented to them had been produced manually or automatically. For each CQ, the questionnaire first presented to the participants the API path manually generated by developers (denoted by "1st"), and then it presents the API path automatically generated by OATAPI (denoted by "2nd"). Participants did not know which paths had been generated by OATAPI.

Results. Regarding the background of the 20 participants, most of them were used to web services and REST APIs, had developed 1–5 web APIs, and were familiar with Semantic Web technologies. As for general questions about REST APIs, most of them knew the correct answer of executing different API calls. As for the last part of the questionnaire, Table 3 summarizes the evaluation results of the API paths shown to the participants. Each row corresponds to the answers of evaluating a CQ and its API paths, and each column header contains the answer options available in the questionnaire.

As shown in the table, participants mostly selected the first and the last option, first and fifth columns respectively. Participants believed that the "1st"

---

[12] The resources, results, and further details of our experiments are available in our repository: https://github.com/oeg-upm/oatapi/tree/main/Evaluation.

API path was built automatically and the "2nd" was generated manually. However, this contradicts real data, as the "1st" are the API paths built by developers and not automatically. As for the last option, participants were not able to determine how the API paths were constructed (indistinguishable). These results indicate that API paths automatically generated with OATAPI are indistinguishable from manually built ones for users.

Table 3. Results of the survey: manually vs automatically built API paths

1st automatically - 2nd manually	2nd automatically - 1st manually	Both manually	Both automatically	Indistinguishable
50%	10%	5%	10%	25%
25%	15%	15%	10%	35%
35%	15%	15%	5%	30%
45%	15%	5%	10%	25%
40%	20%	5%	5%	30%
35%	10%	5%	10%	40%
30%	25%	5%	5%	35%
30%	15%	15%	5%	35%
30%	20%	10%	5%	35%

5.2   Experiment 2

Experiment Settings. For this evaluation, we generated a corpus which contains: 1) a set of CQs and its ontologies (from different domains), 2) API paths manually built, following the steps of our method, for these CQs and ontologies, and 3) API paths built by OATAPI taking these CQs and ontologies as input. In total our final corpus contains 8 ontologies, 50 valid CQs, and 50 API paths. By valid CQs we mean those questions that are not ambiguous, contain terms defined in the ontology, and are not complex. Then, for each CQ we compare the API path manually built and the API path built by OATAPI.

Results. Table 4 summarizes the results of the similarity comparison between the API paths from our corpus. The first column contains the prefixes of the ontologies from our corpus. The values shown in the True Positives column represent the sum of the result of each metric value obtained after executing each comparison. These values are round numbers since we obtained 1 or 0 after evaluating each couple of paths. We evaluated API paths as one token (i.e., if there is a full syntactical match between the paths evaluated).

Table 4. Results of similarity between manually and automatically built API paths.

Ontology	True Positives	API paths evaluated
VGO	15	22
ESCOM	8	8
SWO	0	3
SAREF4ENVI	1	2
NOISE	3	3
PPROC	5	5
ESAIR	3	3
ESBICI	3	4
TOTAL	**38**	**50**

These results show that 38 of the 50 API paths evaluated are similar; (76% accuracy on the API paths). These results indicate that OATAPI is able to generate API paths similar to those built manually by following the steps proposed in our method.

6   Conclusions and Future Work

In this work, we proposed a method designed to extend ontology engineering practices for building APIs. The method provides details on the activities and tasks required to build APIs making use of ontology artefacts, including examples on how to carry them out. From our method, we can conclude that it is possible to generate APIs based on the artefacts created by ontology engineers during the ontology development process (answer to RQ1). We also conclude that making ontology artefacts publicly available is important because these artefacts can be analyzed and reused not only for API generation, but to investigate new solutions for ontology documentation, formalization, testing, or exploitation.

In addition, we developed OATAPI to automate some of the API design tasks of our method. From the evaluation results, we can conclude that it is possible to automatically build API paths that are indistinguishable from those manually generated by application developers (answer to RQ2). We can also conclude that it is possible to automatically generate similar API paths than those manually built following our proposed method. OATAPI allows building API paths that are identical to those generated manually, as evidenced by the high percentage of matches obtained in our second experiment. However, automatic API path generation will depend largely on the conciseness and clarity of the competency questions, and the completeness of the restrictions defined in the ontology.

Although OATAPI allows generating the API paths and SPARQL queries for the CQs, there is still work to be done to improve its functionality. Possible directions include 1) using inference for ontology elements identification to detect those that are not directly defined but are inherited from the parent classes/properties; and 2) enabling synonym detection to ease the ontology elements detection by reusing, for example, the glossary of terms. We aim to move OATAPI from a prototype to a production tool that can be used in real-world scenarios and, as a result, to get wide user feedback, especially from application developers. This feedback will allow us to refine both the method and the tool to continue working on providing resources to facilitate the development of applications that consume KG data.

Acknowledgments. This work was funded by the project Knowledge Spaces: Técnicas y herramientas para la gestión de grafos de conocimientos para dar soporte a espacios de datos (Grant PID2020-118274RB-I00, funded by MCIN/AEI/ 10.13039/501100011033), by the Madrid Government (Comunidad de Madrid-Spain) under the Multiannual Agreement with Universidad Politécnica de Madrid (UPM) in the line Support for R&D projects for Beatriz Galindo researchers, in the context of the V PRICIT (Regional Programme of Research and Technological Innovation), and through the call Research Grants for Young Investigators from UPM.

References

1. Badenes-Olmedo, C., Espinoza-Arias, P., Corcho, O.: R4R: template-based REST API framework for RDF knowledge graphs. In: Proceedings of the ISWC 2021 Posters, Demos and Industry Tracks: From Novel Ideas to Industrial Practice colocated with 20th International Semantic Web Conference, Virtual Conference, 2021. CEUR Workshop Proceedings (2021)
2. Corcho, O., Fernández-López, M., Gómez-Pérez, A.: Methodologies, tools and languages for building ontologies. Where is their meeting point? Data Knowl. Eng. **46**(1), 41–64 (2003)
3. Daga, E., Panziera, L., Pedrinaci, C.: A BASILar approach for building web APIs on top of SPARQL endpoints. In: CEUR Workshop Proceedings, vol. 1359, pp. 22–32 (2015)
4. Daquino, M., Heibi, I., Peroni, S., Shotton, D.: Creating RESTful APIs over SPARQL endpoints using RAMOSE. Semant. Web **23**(2), 195–213 (2022)
5. Espinoza-Arias, P., Garijo, D., Corcho, O.: Crossing the chasm between ontology engineering and application development: a survey. J. Web Semant. **70**, 100655 (2021)
6. Fernández-López, M., Gómez-Pérez, A., Juristo, N.: METHONTOLOGY: from ontological art towards ontological engineering (1997)
7. Fletcher, G., Groth, P., Sequeda, J.: Knowledge scientists: unlocking the data-driven organization. arXiv preprint arXiv:2004.07917 (2020)
8. Garijo, D., Osorio, M.: OBA: an ontology-based framework for creating REST APIs for knowledge graphs. In: Pan, J.Z., et al. (eds.) ISWC 2020. LNCS, vol. 12507, pp. 48–64. Springer, Cham (2020). https://doi.org/10.1007/978-3-030-62466-8_4
9. Jin, B., Sahni, S., Shevat, A.: Designing Web APIs: Building APIs That Developers Love. O'Reilly Media, Inc. (2018)
10. Keet, M.: An Introduction to Ontology Engineering, vol. 1. Maria Keet (2018)
11. Kotis, K.I., Vouros, G.A., Spiliotopoulos, D.: Ontology engineering methodologies for the evolution of living and reused ontologies: status, trends, findings and recommendations. Knowl. Eng. Rev. **35**, E4 (2020)

12. Meroño-Peñuela, A., Hoekstra, R.: grlc makes GitHub taste like linked data APIs. In: Sack, H., Rizzo, G., Steinmetz, N., Mladenić, D., Auer, S., Lange, C. (eds.) ESWC 2016. LNCS, vol. 9989, pp. 342–353. Springer, Cham (2016). https://doi.org/10.1007/978-3-319-47602-5_48

13. Peroni, S.: A simplified agile methodology for ontology development. In: Dragoni, M., Poveda-Villalón, M., Jimenez-Ruiz, E. (eds.) OWLED/ORE - 2016. LNCS, vol. 10161, pp. 55–69. Springer, Cham (2017). https://doi.org/10.1007/978-3-319-54627-8_5

14. Poveda-Villalón, M., Fernández-Izquierdo, A., Fernández-López, M., García-Castro, R.: LOT: an industrial oriented ontology engineering framework. Eng. Appl. Artif. Intell. **111**, 104755 (2022)

15. Salvadori, I., Siqueira, F.: A maturity model for semantic RESTful web APIs. In: 2015 IEEE International Conference on Web Services, pp. 703–710. IEEE (2015)

16. Staab, S., Studer, R., Schnurr, H.P., Sure, Y.: Knowledge processes and ontologies. IEEE Intell. Syst. **16**(1), 26–34 (2001)

17. Suárez-Figueroa, M.C., Gómez-Pérez, A., Fernandez-Lopez, M.: The neon methodology framework: a scenario-based methodology for ontology development. Appl. Ontol. **10**(2), 107–145 (2015)

18. Verborgh, R., Vander Sande, M.: The semantic web identity crisis: in search of the trivialities that never were. Semant. Web J. **11**(1), 19–27 (2020)

MultiAlignNet: Cross-lingual Knowledge Bridges Between Words and Senses

Francesca Grasso[1(✉)] 🆔, Vladimiro Lovera Rulfi[2], and Luigi Di Caro[1] 🆔

[1] University of Turin, Turin, Italy
{fr.grasso,luigi.dicaro}@unito.it
[2] University of Bologna, Bologna, Italy
vladimiro.lovera@unibo.it

Abstract. Numerous NLP applications rely on the accessibility to multilingual, diversified, context-sensitive, and broadly shared lexical semantic information. Standard lexical resources tend to first encode monolithic language-bounded senses which are eventually translated and linked across repositories and languages. In this paper, we propose a novel approach for the representation of lexical-semantic knowledge in - and shared from the origin by - multiple languages, based on the idea of k-Multilingual Concept (MC^k). MC^ks consist of multilingual alignments of semantically equivalent words in k different languages, that are generated through a defined linguistic context and linked via empirically determined semantic relations without the use of any sense disambiguation process. The MC^k model allows to uncover novel layers of lexical knowledge in the form of multifaceted conceptual links between naturally disambiguated sets of words. We first present the conceptualization of the MC^ks, along with the word alignment methodology that generates them. Secondly, we describe a large-scale automatic acquisition of MC^ks in English, Italian and German based on the exploitation of corpora. Finally, we introduce *MultiAlignNet*, an original lexical resource built using the data gathered from the extraction task. Results from both qualitative and quantitative assessments on the generated knowledge demonstrate both the quality and the novelty of the proposed model.

Keywords: Lexical Semantics · Multilingual alignments

1 Introduction

The exploitation of lexical resources constitutes a key issue for several Natural Language Processing tasks and applications. Many existing resources, such as WordNet [30], usually encode language-bounded lexical knowledge in the form of *word senses*, i.e., dictionary-oriented definitions of lexical entries which are linked and put in context through lexical-semantic relations. These relations, being only of a paradigmatic nature, are characterized by a sharing of similar defining properties between the words and a requirement that the items belong to the same syntactic category [32]. The fine-grained structure of such resources

© The Author(s), under exclusive license to Springer Nature Switzerland AG 2022
O. Corcho et al. (Eds.): EKAW 2022, LNAI 13514, pp. 36–50, 2022.
https://doi.org/10.1007/978-3-031-17105-5_3

and the lack of syntagmatic associations, while allowing a high systematization of the linguistic data, determines an artificial abstraction that does not always reflect empirical reality. This is mainly due to the lack of a meaning encoding system capable of representing concepts in a flexible way [35].

Word Sense Disambiguation (WSD) is the task of determining the context-consistent meaning of a word from among all its possible senses by drawing from a sense repository [33]. Sense repositories may vary in terms of generality (from top-level and general purposes up to domain-specific ones) and completeness. WordNet is currently one of the most commonly adopted, with counterparts in other languages [5] and links with other resources, e.g. BabelNet [34]. While many works focused on raising the state-of-the-art performance, the improvement still stops at 81% of F-score when using WordNet as sense inventory [3,26]. This is due to the difficulty to perform disambiguation, which constitutes one of the more complex and elusive processes of the semantic landscape even in human-to-human dialogues [13,37]. Current state-of-the-art approaches are mainly devoted to create or link repositories rather than clustering existing senses. In this paper we propose a different approach, providing a natively cross-lingual view of the problem.

As is known, lexical ambiguity is a natural property of semantic systems which, however, mutates from language to language. Therefore, it may decrease when putting lexical items in reciprocal relation, i.e., when aligned. While a given language may provide only a single disambiguation context for a word, the use of parallel languages may indeed help further restrict word sense variability [21]. For example, the concept of *"discharge from an office or position"* may be encoded into the English verb form *"to fire"* which is however highly ambiguous, counting twelve different verbal senses in WordNet. The same concept is expressed by another polysemous term in Italian, i.e. *"licenziare"*. However, the words *fire - licenziare* when associated with each other represent a bilingual encoding of that single concept which naturally avoids ambiguity, given that there are no other meanings that the two words may share. Thus, translations of a target word into one or more languages provide it a disambiguation context and may serve as sense labels [27]. Many works [1,8,10,12,27], have already shown the advantages of multilingual word alignments to perform Word Sense Disambiguation, although dwelling on the exploitation of either parallel corpora or multilingual wordnets, i.e., on already existing and pre-determined cross-lingual lexical material. In this work, we propose to leverage this property of languages for a broader purpose.

First, we propose a novel lexical-semantic encoding model bridging between words and senses called *k-Multilingual Concept* (MC^k), based on the above-mentioned cross-lingual alignment in k different languages. As a second contribution, we present a large-scale automatic acquisition of MC^ks from several corpora in three languages (English, Italian, and German). This model enables the encoding of varied layers of lexical knowledge, in terms of both syntagmatic and paradigmatic relations, providing networks of diversified conceptual links between words in - and shared by - different languages. Through the proposed method we extracted a total of $21,514$ trilingual alignments belonging to three different types of Part-of-Speech tags (nouns, modifiers and verbs) for more than 1,047 input WordNet synsets. As final contribution, we publicly release a resource, called *MultiAligNet*, in two different versions, i.e. in *i)* vectorial and *ii)* graph-based forms. Finally, we evaluate the resource through both qualitative and quantitative assessments, demonstrating *i)* the high quality of the extracted multilingual alignments, *ii)* the novelty of the uncovered lexical semantic relations, and *iii)* the natural (rather than artificial) disambiguation power of the proposed multilingual approach.

2   Related Work

The problem of identifying the correct meaning of words depending on the context of occurrence represents one of the oldest tasks in the field of Natural Language Processing. The process of Word Sense Disambiguation hides a wide range of complexities, such that even after decades of technological advancement the current state of the art is still far from reaching more-than-good accuracy levels [26]. Many studies have already proved the advantages of a cross-lingual approach to Word Sense Disambiguation [1,8,10,12]. The use of translations of a given word as sense labels avoid the need for manually created sense-tagged corpora and sense inventories. Moreover, a cross-lingual approach deals with the sense granularity problem: finer sense distinctions became truly relevant as far as they get lexicalized into different translations of the word [27]. However, existing works usually exploit either parallel texts or multilingual Wordnets, therefore relying on a intrinsically limited number of de-facto already built alignments.

Standard ways to encode lexical meaning are often based on explicit links between *words* and their possible *senses*, whereas words/senses are connected via paradigmatic relations (e.g., hypernymy, synonymy, antonymy, etc.), as in Word-Net [30] and BabelNet [34]. Extensions of these resources also include Common-Sense Knowledge (CSK), which refers to some (to a certain extent) widely-accepted and shared information. CSK describes the kind of general knowledge material that humans use to define, differentiate and reason about the conceptualizations they have in mind. ConceptNet [42] is one of the largest CSK resources, collecting and automatically integrating data starting from the original MIT Open Mind Common Sense project[1]. However, terms in ConceptNet are not disambiguated. Property norms [11,28] represent a similar kind of resource, which is more focused on cognitive and perception-based aspects of word meaning. Norms, in contrast with ConceptNet, are based on semantic features empirically-constructed via questionnaires producing lexical (often ambiguous) labels associated with target concepts, without any systematic methodology of knowledge collection and encoding. An emerging and extremely impactful approach to lexical semantics has been adopted by corpus-based and data-driven studies and technologies, which led to the creation of numeric (vectorial) encoding of lexical knowledge. This method is all centered on Harris' distributional assumption [17],

---

[1] https://www.media.mit.edu/.

i.e. words that occur in the same contexts tend to have similar meanings. Well-known models include word embeddings [4,29,36], sense embeddings [19,20,25], and contextualized embeddings [39]. However, the relations holding between vector representations are not typed, nor are they organized systematically.

3 k-Multilingual Concepts

In this paper, we first propose the idea of *k-Multilingual Concept* (hereinafter MC^k), which consists of a concatenation of k lexical items referring to a single concept in k different languages. A MC^k can be described as a *pseudoword*, in line with the proposals put forward by [15] and [40], i.e., artificially-created words that can be used for different purposes (e.g., for the evaluation of Word Sense Induction systems [38]). In this instance, MC^ks are pseudowords that result from (and consist of) the alignment of multilingual, semantically equivalent lexical forms of a given concept. For example, if we consider the concept "*cat*" (as"*domestic cat*"), its $MC^{EN,IT}$ for the two languages English and Italian would be:

$$cat^{EN} \oplus gatto^{IT}$$

where the symbol $\oplus$ represents a simple concatenation operator. Similarly, we may extend the string by including other languages, adding e.g. a German equivalent word form. We would therefore obtain the following $MC^{EN,IT,DE}$:

$$cat^{EN} \oplus gatto^{IT} \oplus Katze^{DE}$$

A single MC^k is thus composed of k lexical forms, each one being linked to a specific language. However, the idea of a MC^k also presupposes that each of the k languages may have from zero to multiple lexicalizations of a given concept. The latter case would involve a synonymical set of words, whereas the former denotes what is referred to as lexical gap, i.e., concepts that lexicalize in one language but not in another. For example, the German reflexive verb *fremdschämen* in both Italian and English needs to be expressed with a periphrasis such as "*to feel embarrassed for someone*", since there is no lexical item with an equivalent meaning in the lexicons of either languages.

3.1 Lexical Gaps

Lexicalization is one of the linguistic devices available in natural languages for the integration of an item into the lexicon. This phenomenon typically involves a previously morphologically complex word that starts to acquire semantic and functional autonomy and behave as a single and independent lexical unit [43]. Being both a semantic notion and a process, it is gradient rather than categorical. Therefore, there can be different degrees of lexicalization. For example, the concept $\{leisure^{EN}, Freizeit^{DE}\}$ must be expressed in Italian through the multi-word expression *tempo libero*^{IT}. Despite being formed by two words, this expression nevertheless displays the same morphosyntactic and functional properties of the corresponding lexical forms in English and German. Thus, while *fremdschämen* is fully unlexicalized in Italian and English and generates a lexical gap, many lexical units such as *tempo libero^{IT}* or, e.g., English phrasal verbs represent lexical entries[2] albeit being slightly less-lexicalized than single-word units. Whenever the inventory of lexemes of a language does not include the full lexicalization of a given concept, such a lexical gap may create an empty value within a MC^k. This would be the case of *fremdschämen* or, e.g., of the Italian word *abbiocco* - which specifically denotes a feel of sleepiness caused by the digestion of an heavy meal. Thus, we will have:

$$\{\}^{EN} \oplus abbiocco^{IT} \oplus \{\}^{DE}$$

as $MC^{EN,IT,DE}$ associated with this concept. The idea of *"move body upright from sitting or lying"*, instead, will be regularly encoded into the following $MC^{EN,IT,DE}$:

$$stand\ up^{EN} \oplus alzarsi^{IT} \oplus aufstehen^{DE}$$

3.2   Synonymous Words

A language may encode identical or similar semantic content into multiple word forms, causing instances of synonymy[3]. This will lead to a plurality of coordinated terms within the MC^k for a single concept. For example, if we only consider the English synonymical word forms *bike* and *bicycle*, we would have:

$$\{bike, bicycle\}^{EN} \oplus bicicletta^{IT} \oplus Fahrrad^{DE}$$

as $MC^{EN,IT,DE}$ associated with that single meaning[4].

3.3   Polysemous Words

Among the complex peculiarities of natural languages, that of polysemy (or *semantic ambiguity*) represents notoriously a challenging phenomenon for Natural Language Processing. Polysemy refers to the capacity for a word to convey multiple meanings, whereas the process of identification of its context-sensitive meaning is called disambiguation. However, each language features its own peculiar semantic system which, in turn, employs different formal encoding strategies. Therefore, by exploiting the different semantic (i.e. polysemous) behaviours of lexical items it is possible to disambiguate a given word by means of its semantic counterpart in another language.

---

[2] Therefore they are formally included in dictionaries, being considered as part of the lexicon by lexicographers.

[3] Yet synonymy, as a rule, is not complete equivalence - as we are reminded by [22].

[4] The same would apply for Italian and German synonyms for the concept *bicycle*.

The presented idea of MC^k is meant to represent a key instrument in this respect, since it is composed of a set of semantically equivalent lexical items that provide a quasi-monosemic (i.e. disambiguated) multilingual alignment. By providing a MC^k a context, or, more accurately, when a MC^k is generated through a defined linguistic context, their members will be indeed assigned a context-consistent meaning. Therefore, the MC^k will pinpoint a specific and unique concept. Finally, starting from the proven practice of leveraging multilingual word alignments to perform word disambiguation, we propose a novel methodology for automatically build them on a large scale without relying on already provided translations.

In the next section we will describe in detail the multilingual alignment mechanism that generates the MC^ks. This methodology, taken directly from [16], underpins the implementation of the MC^ks extraction as described thereafter.

4 Alignment Methodology

In this section, we present the alignment methodology used to automatically extract k-Multilingual Concepts from language-specific corpora.

4.1 Method and Languages Involved

As already performed in [16] we use three different languages in order to illustrate the building process of the multilingual resource. Thus, three European languages are involved in our work: English, German and Italian. The choice fell on these primarily because we are proficient in them, therefore we are able to properly handle and interpret the data. Furthermore, due to the very nature of the methodology, it was advisable to select a set of languages featuring a certain level of similarity in terms of shared lexical-semantic material. At the present stage, the alignment mechanism can be indeed effective and the results appreciable as long as the lexical-semantic systems of the languages involved reflect compatible cultural-linguistic backgrounds. A basic example will now help introduce the multilingual alignment mechanism. Consider the concept *"wool"* (as *"textile fiber obtained from sheep and other animals"*) and the tree word forms $\{wool^{EN}, lana^{IT}, Wolle^{DE}\}$, constituting the following $MC^{EN,IT,DE}$:

$$wool^{EN} \oplus lana^{IT} \oplus Wolle^{DE}$$

The so conceived *head* concept represents our starting point from which a linguistic context will be generated. Hence, we may represent it also as:

$$MC^{EN,IT,DE}_{wool-textile\ fiber}$$

For each of the three word forms that compose the $MC^{EN,IT,DE}$ *head* we retrieve a set of semantically related words of different types (nouns, modifiers, verbs) in terms of paradigmatic (e.g. synonyms) and syntagmatic (e.g. co-occurrences) relations. We thus obtain three different lists of *head*-related

Table 1. Unordered lists of single-language related words for $MC^{EN,IT,DE}_{wool-textile\ fiber}$.

wool^{EN}	lana^{IT}	Wolle^{DE}
sheep	cotone	Schal
cotton	Biella	spinnen
synthetic	sintetica	Baumwolle
spin	sciarpa	Rudolf
scarf	pecora	synthetisch
mitten	filare	Schafe

words, one for each of the three languages. Table 1 provides a small excerpt of such unordered lists.

The retrieved terms in the lists may be still ambiguous, since they are related to a word form rather than to a contextually defined concept. Thus, the lexical data in the lists are subsequently compared and filtered by means of a translation step, in order to select only the semantic items that occur in all the lists, i.e., those shared by the three languages. The resulting words are thus aligned with their semantic counterparts, as shown in Table 2.

Table 2. Examples of aligned concept-related words for $MC^{EN,IT,DE}_{wool-textile\ fiber}$.

wool^{EN}		lana^{IT}		Wolle^{DE}
sheep	⊕	pecora	⊕	Schafe
cotton	⊕	cotone	⊕	Baumwolle
synthetic	⊕	sintetica	⊕	synthetisch
spin	⊕	filare	⊕	spinnen
scarf	⊕	sciarpa	⊕	Schal

As can be noted, by combining, e.g., the lexical form *to spin* with the Italian word *filare* and the German *spinnen* - which, among others, encode one of the possible senses of *spin* - we would obtain the following $MC^{EN,IT,DE}$:

$$spin^{EN} \oplus filare^{IT} \oplus spinnen^{DE}$$

Once aligned, the three previously polysemous lexical forms constitute a $MC^{EN,IT,DE}$ that refers to a specific and unique conceptualization, i.e., *"turn fibers into thread"*. The resulting list of $MC^{EN,IT,DE}$ for the head concept $MC^{EN,IT,DE}_{wool-textile\ fiber}$ provides an encoding of lexical knowledge linked to the seed concept which is i) *unbiased*, since the filtering step enables to avoid language-bounded material by including only items that are shared by all three languages; ii) *diversified*, since it consist of both paradigmatic and syntagmatic lexical relations for three different POS.

4.2 Automatic Extraction of MC^ks

We built a data ingestion process that automatically outputs MC^ks, using as mentioned above $k=3$ languages: English (EN), Italian (IT) and German (DE). To start an automatic MC^k extraction process for a generic concept C the first requirement is to have a seed, i.e., a MC^k *head* that is constituted by k word forms representing C, one for each language. Since a generic concept C may present language-related issues (e.g. lexical gaps - see Sect. 3.1), we retrieve MC^k *heads* directly from BabelNet synsets. In particular, given a BabelNet synset for a concept C, we select a maximum of 3 *high-quality* lexicalizations[5] for each language. If BabelNet does not provide at least one high quality lexicalization for each language, we rely on Open Multilingual Wordnet project [6] to look for English and Italian lexicalizations and OdeNet [41] for German ones, while Collaborative InterLingual Index (CILI) [7] serves as a link between the two to retrieve the shared synset. The obtained word forms in the three languages will constitute the MC^k *head* around which the procedure will autonomously extract the multilingual knowledge around C.

Once the MC^k *head* has been formed, we use Sketch Engine [24], a corpus management engine, to obtain lists of words related to each single word form that makes up the MC^k *head*, as shown in the example in Table 1. We employ three families of non-semantically annotated large corpora to search for related words in the three languages: the TenTen corpora containing 10+ billion words of generic web content [23], the TJSI corpora composed of news articles [44][6] and the EUR-Lex legal corpora [2]. Then, we merge the retrieved related words in the three target languages obtaining three lists (hereinafter *EN*-list, *IT*-list and *DE*-list), each divided into four categories: *i)* similar nouns, *ii)* co-occurring nouns, *iii)* co-occurring adjectives and *iv)* co-occurring verbs. Finally, we assign a weight to each related word by directly importing the built-in scores of Sketch Engine tools, that are based on the Dice coefficient, as detailed in [24].

To obtain the the MC^ks alignments like those shown in Table 2 we search for cross-match translations using the PanLex API[7], which is focused on words rather than on sentences, and the Google Translate API[8]. Specifically, we take each related word, category by category, from the *EN*-list and query the API to get their possible translations into Italian, ordered by confidence. If we find a match between such translations and a related word in the *IT*-list of equal category, we form a pair $<rw^{EN}, rw^{IT}>$. Once all possible pairs have been identified, we repeat the procedure starting from all rw^{EN}s to find matches within the *DE*-list of the same category, thus obtaining triplets $<rw^{EN}, rw^{IT}, rw^{DE}>$. A final verification is performed by testing the correct correspondence between each $<rw^{IT}, rw^{DE}>$ pair, through the same cross-match translation process. If this

---

[5] BabelNet high-quality lexicalizations are those word forms that are not marked as resulting from an automatic translation.

[6] TJSI versions used: English (60+ billion words), Italian (8.4+ billion words), German (6.9+ billion words).

[7] https://dev.panlex.org/api/.

[8] https://cloud.google.com/translate.

step fails, the whole triplet will be marked as *weak*. Otherwise, the successful alignment will be considered as *strong* and will constitute a $MC^{EN,IT,DE}$. We finally assign a score to each $MC^{EN,IT,DE}$ by averaging the SketchEngine scores of the three related words.

As last step, we associate BabelNet synsets (always those directly linked to WordNet synsets, if present) and WordNet synsets to the alignments. Specifically, we find the n synsets that have all the given three word forms in the three languages. One of the following three cases may hence occur: *i)* $n = 1$, meaning that the $MC^{EN,IT,DE}$ corresponds to a completely disambiguated concept; *ii)* $n > 1$, when multiple synsets may be associated with a single $<rw^{EN}, rw^{IT}, rw^{DE}>$ triplet; *iii)* $n = 0$, in case no existing BabelNet synset or WordNet synset actually connects the three word forms. It is interesting to note that the last two cases cover different situations, such as a missing synset econding a specific concept ($n = 0$, e.g. significant for sense induction) or overlapping synsets ($n > 1$, e.g. useful for sense clustering).

5   The *MultiAligNet* Resource

The k-Multilingual Concept model and the automatic extraction method we developed allowed us to create an original lexical-semantic resource, which we refer to as *MultiAligNet*. To date, the resource is publicly available[9] and contains the extracted knowledge referring to 1047 synsets that we used as *heads*, which corresponds to a total of 21514 automatically-built MC^ks over the three languages. Future updates will be made available within the same repository. The selection of *head* concepts has been performed carefully. First, we manually selected 100 concepts by inspecting basic vocabularies of each of the three languages[10], covering different semantic categories and characteristics such as the degrees of polysemy and abstractness. Then we automatically retrieved the 750 most frequent and 200 rare concepts in SemCor [31], one of the most used sense-annotated corpora to train supervised WSD systems. Finally, we randomly-picked a set of polysemous words referring to more than 50 synsets in total. The *MultiAligNet* resource is available in two different formats, as described below.

5.1   Distributional Representation

Our resource can be displayed through a vectorial representation of the k-Multilingual Concepts. In particular, synsets are represented as vectors whose dimensions point to the synsets linked to the alignments (see Sect. 4.2 for details). Such distributional version of the resource is different from standard word- and

---

[9] https://github.com/vloverar/multialignet.
[10] For EN: iWebCorpus, The Oxford Dictionary https://www.english-corpora.org/ iweb, https://www.oxfordlearnersdictionaries.com/wordlists/oxford3000-5000; for IT: *NvdB* https://www.dropbox.com/s/mkcyo53m15ktbnp/nuovovocabolariodibase. pdf; for DE: [45].

sense-embedding technologies, since features are conceptual (being connected to real synsets). This is similar to what happens with Explicit Semantic Analysis (ESA) [14], Salient Semantic Analysis (SSA) [18] and others [9]. This version may be employed in semantic similarity tasks and, generally, in the context of Explainable AI research.

5.2   Knowledge Graph

Similarly to other lexical-semantic resources, our model reflects a deep interconnection of term- and concept-based items, which makes it well-suited for a graph-based knowledge encoding. We provide a knowledge graph relying on the Neo4j[11] database open technologies and libraries. In the graph model we employ four types of nodes, namely *i) word*-nodes, *ii) babel synset*-nodes, *iii) wordnet synset*-nodes and *iv) align*-nodes (further typed with POS tags). While the first

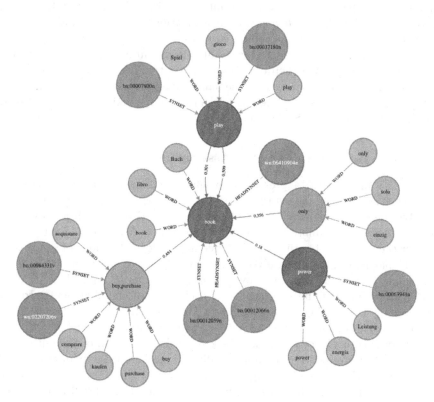

Fig. 1. Illustrative excerpt of *MultiAligNet* graph around the $MC^{EN,IT,DE}_{book-written\ work}$ head. Red, pink and green circles represent *align*-nodes for nouns, verbs and adjectives respectively (for space requirements, only the English word forms are displayed). Beige, blue and orange ones represent *word*-, *babel synset*- and *wordnet synset*-nodes.

---

[11] https://neo4j.com.

three enable standard access features for words- and synsets-centered queries (as in WordNet and BabelNet), *align*-nodes represent a novel type of information, specifically hinged on the MC^k multilingual concatenations of terms. The released *MultiAligNet* knowledge graph contains 72,469 nodes, interconnected by 387,273 relations. Figure 1 shows an excerpt of the graph around the $MC^{EN,IT,DE}_{book-written\ work}$ head.

6 Extraction Results and Evaluation

Starting from our selected concepts (1,047 *heads*), we automatically extracted 21,514 multilingual alignments (MC^ks). Among them, 9,007 (41.86%) do not present any available linking to either WordNet or BabelNet synsets (for the latter, considering only the *high quality* lexicalizations) whereas 1,045 have an available linking only to *low-quality* lexicalization in BabelNet. Finally, 7,962 triplets (37.01%) present no available linking to either WordNet or BabelNet, considering both *high-* and *low quality* lexicalizations. This latter data refers to totally novel lexical knowledge compared to the two reference resources.

In this section, we first report the results of a qualitative assessment of such generated knowledge. We then outline a quantitative evaluation reflecting the impact of MC^ks in uncovering novel semantic relations with respect to a state-of-the-art existing repository (i.e. BabelNet) without making use of any Word Sense Disambiguation (WSD) system.

6.1 MC^ks Novelty and Quality Assessment

7,962 MC^ks out of 21,514 present no available linking to either WordNet or Babelnet synsets. This means that the system managed to retrieve novel lexical knowledge quantifiable as 7,962 alignments related to 1,047 head concepts. We then manually evaluated the quality of these new MC^ks in order to assess whether they consist of actually valid three-lingual lexicalizations of single concepts. In particular, we manually checked a randomized subset of 250 triplets. The manual check was performed by assessing the semantic equivalence of each MC^k, thus validating the translations of each word of the alignment into the other two by using bilingual dictionaries[12]. We assessed both translation directions for each word pair ($<rw^{EN}, rw^{IT}>$; $<rw^{EN}, rw^{DE}>$; $<rw^{DE}, rw^{IT}>$). The semantic equivalence assessment task showed that a total of 235 out of 250 MC^k (93.6%) were indeed accurate. Finally, we measured the amount of novel connections retrieved by MultiAligNet with respect to the BabelNet knowledge graph. Interestingly, 264,813 links between alignments (out of 290,730) are not present in BabelNet.

---

[12] The annotator who performed the evaluation is however a native Italian speaker with a minimum of C1 both English and German proficiency level. Therefore, the evaluation is assured by a solid accuracy.

6.2 MC^ks Disambiguation Power

The MC^k model enables a peculiar encoding of lexical knowledge which lies between the high polysemy of words and the static nature of predefined word senses. Therefore, we aim to concretely measure to what extent MC^ks can reduce single-language word ambiguity without relying on any WSD method. Hence, for each polysemous word w^L in a given language L, we can count its possible senses $ns(w^L) \geq 2$, as well as the resulting senses linked to the k-multilingual concept $ns(MC^k_{w^L})$. Note that $ns(w^L)$ is always greater than or equal to $ns(MC^k_{w^L})$. We can compute a disambiguation power (dp) index for a single word w^L as follows:

$$dp(w^L, MC^k_{w^L}) = \frac{ns(w^L) - max(1, ns(MC^k_{w^L}))}{ns(w^L) - 1}$$

Note that since MC^ks may not be linked to any synset (as mentioned in Sect. 4.2), the max function forces to 1 the value of the subtrahend. The range of the dp is $[0, 1]$ where 0 means no disambiguation and 1 maximum disambiguation (this latter case occurs whenever all senses $ns(w^L)$ got reduced to a single MC^k sense (i.e. $ns(MC^k_{w^L}) = 1$)). In order to obtain an overall MC^k dp-index for a set of target words in a language L, we can compute an average score as follows:

$$dp^L = \frac{1}{|w^L|} \sum_{\forall w^L} dp(w^L, MC^k_{w^L})$$

Table 3 shows the dp index for the three languages. Impressively, MC^ks considerably reduced single-language word ambiguity in all three languages. In particular, for the *EN*- and *IT*-ambiguous lexical entries, the proposed alignment was able to reduce their polysemy by 85%. This demonstrates the high potential of the MC^k model in encoding mostly-unambiguous lexical knowledge without relying on fixed sense repositories.

Table 3. Disambiguation power (dp) index for the three languages *EN*, *IT*, *DE*.

Language	N. of ambiguous words	dp-index
EN	9480	0.851
IT	7395	0.852
DE	4866	0.756

7 Conclusion and Future Work

In this paper, we proposed a novel encoding method for the representation of lexical-semantic knowledge based on the idea of k-Multilingual Concept (MC^k). The developed methodology allows the automatic alignment of semantically equivalent words in k different languages as occurring in a determined linguistic context. The resulting alignments result in a cross-lingual encoding of unbiased and multifaceted lexical knowledge, in terms of empirically determined conceptual links consisting of syntagmatic and paradigmatic lexical relations.

We then released *MultiAligNet*, an original resource containing, to date, more than 21k automatically-extracted MC^ks on a heterogeneous selection of concepts in English, Italian and German. We thus evaluated the resource by means of both qualitative and quantitative assessments on the data retrieved. Results demonstrate the validity of the method concerning its ability to retrieve (i) unbiased lexical knowledge (ii) diversified lexical relations (iii) novel lexical material as compared to existing resources (BabelNet and WordNet). Finally, the proposed model enabled a natural (multilingual) disambiguation mechanism for words without the help of sense repositories or parallel texts. In future work, we aim to continuously extend the resource by covering more concepts and languages, fostering novel research on different tasks such as enrichment, disambiguation and induction of senses in existing repositories.

References

1. Apidianaki, M.: LIMSI: cross-lingual word sense disambiguation using translation sense clustering. In: Second Joint Conference on Lexical and Computational Semantics (*SEM), Volume 2: Proceedings of the Seventh International Workshop on Semantic Evaluation (SemEval 2013), pp. 178–182. Association for Computational Linguistics, Atlanta (2013). https://aclanthology.org/S13-2032
2. Baisa, V., et al.: European union language resources in sketch engine. In: Proceedings of the Tenth International Conference on Language Resources and Evaluation (LREC'16), pp. 2799–2803 (2016)
3. Barba, E., Procopio, L., Navigli, R.: Consec: Word sense disambiguation as continuous sense comprehension. In: Proceedings of the 2021 Conference on Empirical Methods in Natural Language Processing, pp. 1492–1503 (2021)
4. Bojanowski, P., Grave, E., Joulin, A., Mikolov, T.: Enriching word vectors with subword information. arXiv preprint arXiv:1607.04606 (2016)
5. Bond, F., Foster, R.: Linking and extending an open multilingual wordnet. In: Proceedings of the 51st Annual Meeting of the Association for Computational Linguistics (Volume 1: Long Papers), pp. 1352–1362 (2013)
6. Bond, F., Foster, R.: Linking and extending an open multilingual Wordnet. In: Proceedings of the 51st Annual Meeting of the Association for Computational Linguistics (Volume 1: Long Papers), pp. 1352–1362. Association for Computational Linguistics, Sofia (2013). https://aclanthology.org/P13-1133
7. Bond, F., Vossen, P., McCrae, J., Fellbaum, C.: CILI: the collaborative interlingual index. In: Proceedings of the 8th Global WordNet Conference (GWC), pp. 50–57. Global Wordnet Association, Bucharest (2016)
8. Brown, P.F., Della Pietra, S.A., Della Pietra, V.J., Mercer, R.L.: Word-sense disambiguation using statistical methods. In: 29th Annual Meeting of the Association for Computational Linguistics, pp. 264–270. Association for Computational Linguistics, Berkeley (1991)
9. Camacho-Collados, J., Pilehvar, M.T., Navigli, R.: Nasari: integrating explicit knowledge and corpus statistics for a multilingual representation of concepts and entities. Artif. Intell. **240**, 36–64 (2016)

10. Chan, Y.S., Ng, H.T.: Scaling up word sense disambiguation via parallel texts. In: Proceedings of the 20th National Conference on Artificial Intelligence (AAAI'05) - Volume 3, pp. 1037–1042. AAAI Press (2005)
11. Devereux, B.J., Tyler, L.K., Geertzen, J., Randall, B.: The CSLB concept property norms. Behav. Res. Methods **46**(4), 1119–1127 (2014)
12. Diab, M.T., Resnik, P.: Word Sense Disambiguation within a Multilingual Framework. Ph.D. thesis, USA, aAI3115805 (2003)
13. Edmonds, P., Kilgarriff, A.: Introduction to the special issue on evaluating word sense disambiguation systems. Nat. Lang. Eng. **8**(4), 279–291 (2002)
14. Gabrilovich, E., Markovitch, S., et al.: Computing semantic relatedness using wikipedia-based explicit semantic analysis. In: IJcAI, vol. 7, pp. 1606–1611 (2007)
15. Gale, W.A., Church, K.W., Yarowsky, D.: Work on statistical methods for word sense disambiguation. In: Working Notes of the AAAI Fall Symposium on Probabilistic Approaches to Natural Language, vol. 54, p. 60 (1992)
16. Grasso, F., Di Caro, L.: A methodology for large-scale, disambiguated and unbiased lexical knowledge acquisition based on multilingual word alignment. In: Fersini, E., Passarotti, M., Patti, V. (eds.) Proceedings of the Eighth Italian Conference on Computational Linguistics, CLiC-it 2021, Milan, Italy, 26–28 January 2022. CEUR Workshop Proceedings, vol. 3033. CEUR-WS.org (2021)
17. Harris, Z.S.: Distributional structure. Word **10**(2–3), 146–162 (1954)
18. Hassan, S.H., Mihalcea, R.: Semantic relatedness using salient semantic analysis. In: Twenty-Fifth AAAI Conference on Artificial Intelligence (2011)
19. Huang, E.H., Socher, R., Manning, C.D., Ng, A.Y.: Improving word representations via global context and multiple word prototypes. In: Proceedings of ACL, pp. 873–882 (2012)
20. Iacobacci, I., Pilehvar, M.T., Navigli, R.: SensEmbed: learning sense embeddings for word and relational similarity. In: Proceedings of ACL, pp. 95–105 (2015)
21. Ion, R., Tufis, D.: Multilingual word sense disambiguation using aligned wordnets. Romanian J. Inf. Sci. Technol. **7**, 183–200 (2004)
22. Jakobson, R.: 14. On Linguistic Aspects of Translation, pp. 144–151. University of Chicago Press (2012)
23. Jakubíček, M., Kilgarriff, A., Kovář, V., Rychlý, P., Suchomel, V.: The tenten corpus family. In: 7th International Corpus Linguistics Conference CL, pp. 125–127 (2013)
24. Kilgarriff, A., et al.: The sketch engine: ten years on. Lexicography **1**(1), 7–36 (2014)
25. Kumar, S., Jat, S., Saxena, K., Talukdar, P.: Zero-shot word sense disambiguation using sense definition embeddings. In: Proceedings of the 57th Annual Meeting of the Association for Computational Linguistics, pp. 5670–5681 (2019)
26. Lacerra, C., Bevilacqua, M., Pasini, T., Navigli, R.: CSI: a coarse sense inventory for 85% word sense disambiguation. In: Proceedings of the AAAI Conference on Artificial Intelligence, vol. 34, pp. 8123–8130 (2020)
27. Lefever, E., Hoste, V.: SemEval-2013 task 10: cross-lingual word sense disambiguation. In: Second Joint Conference on Lexical and Computational Semantics (*SEM), Volume 2: Proceedings of the Seventh International Workshop on Semantic Evaluation (SemEval 2013), pp. 158–166. Association for Computational Linguistics, Atlanta (2013). https://aclanthology.org/S13-2029
28. McRae, K., Cree, G.S., Seidenberg, M.S., McNorgan, C.: Semantic feature production norms for a large set of living and nonliving things. Behav. R. M. **37**(4), 547–559 (2005)

29. Mikolov, T., Sutskever, I., Chen, K., Corrado, G.S., Dean, J.: Distributed representations of words and phrases and their compositionality. In: Advances in Neural Information Processing Systems, pp. 3111–3119 (2013)
30. Miller, G.A.: Wordnet: a lexical database for English. Commun. ACM **38**(11), 39–41 (1995)
31. Miller, G.A., Chodorow, M., Landes, S., Leacock, C., Thomas, R.G.: Using a semantic concordance for sense identification. In: Human Language Technology: Proceedings of a Workshop held at Plainsboro, New Jersey, 8–11 March 1994 (1994)
32. Morris, J., Hirst, G.: Non-classical lexical semantic relations. In: Proceedings of the Computational Lexical Semantics Workshop at HLT-NAACL 2004, pp. 46–51. Association for Computational Linguistics, Boston (2004). https://aclanthology.org/W04-2607
33. Navigli, R.: Word sense disambiguation: a survey. ACM Comput. Surv. **41**(2), 1–69 (2009)
34. Navigli, R., Ponzetto, S.P.: BabelNet: building a very large multilingual semantic network. In: Proceedings of ACL, pp. 216–225. Association for Computational Linguistics (2010)
35. Palmer, M., Dang, H.T., Fellbaum, C.: Making fine-grained and coarse-grained sense distinctions, both manually and automatically. Nat. Lan. Eng. **13**(02), 137–163 (2007)
36. Pennington, J., Socher, R., Manning, C.D.: Glove: Global vectors for word representation. In: EMNLP, vol. 14, pp. 1532–1543 (2014)
37. Petricca, P.: SEMANTICA. Forme, Modelli, Problemi (2019)
38. Pilehvar, M.T., Navigli, R.: A large-scale pseudoword-based evaluation framework for state-of-the-art word sense disambiguation. Comput. Linguist. **40**(4), 837–881 (2014)
39. Scarlini, B., Pasini, T., Navigli, R.: SensEmBERT: context-enhanced sense embeddings for multilingual word sense disambiguation. In: Proceedings of the 34th Conference on Artificial Intelligence. Association for the Advancement of Artificial Intelligence (2020)
40. Schütze, H.: Dimensions of meaning. In: SC, pp. 787–796 (1992)
41. Siegel, M., Bond, F.: OdeNet: compiling a GermanWordNet from other resources. In: Proceedings of the 11th Global Wordnet Conference, pp. 192–198. Global Wordnet Association, University of South Africa (UNISA) (2021). https://aclanthology.org/2021.gwc-1.22
42. Speer, R., Chin, J., Havasi, C.: Conceptnet 5.5: an open multilingual graph of general knowledge (2017)
43. Thomas, C.: Lexicalization in Generative Morphology and Conceptual Structure, pp. 45–65. Edinburgh University Press (2013)
44. Trampuš, M., Novak, B.: Internals of an aggregated web news feed. In: Proceedings of 15th Multiconference on Information Society, pp. 221–224 (2012)
45. Tschirner, E.: Deutsch nach Themen: Grund-und Aufbauwortschatz: Deutsch als remdsprache nach Themen-Lernwörterbuch. Cornelsen, Berlin (2016)

Question Answering with Additive Restrictive Training (QuAART): Question Answering for the Rapid Development of New Knowledge Extraction Pipelines

Corey A. Harper[1,2]([✉]), Ron Daniel Jr.[1], and Paul Groth[2]

[1] Elsevier Labs, Suite 800, 230 Park Avenue, New York, NY 10169, USA
{c.harper,r.daniel}@elsevier.com
[2] University of Amsterdam, Postbus 94323, 1090 GH Amsterdam, The Netherlands
{c.a.harper,p.t.groth}@uva.nl

Abstract. Numerous studies have explored the use of language models and question answering techniques for knowledge extraction. In most cases, these models are trained on data specific to the new task at hand. We hypothesize that using models trained only on generic question answering data (e.g. SQuAD) is a good starting point for domain specific entity extraction. We test this hypothesis, and explore whether the addition of small amounts of training data can help lift model performance. We pay special attention to the use of null answers and unanswerable questions to optimize performance. To our knowledge, no studies have been done to evaluate the effectiveness of this technique. We do so for an end-to-end entity mention detection and entity typing task on HAnDS and FIGER, two common evaluation datasets for fine grained entity recognition. We focus on fine-grained entity recognition because it is challenging scenario, and because the long tail of types in this task highlights the need for entity extraction systems that can deal with new domains and types. To our knowledge, we are the first system beyond those presented in the original FIGER and HAnDS papers to tackle the task in an end-to-end fashion. Using an extremely small sample from the distantly-supervised HAnDS training data – 0.0015%, or less than 500 passages randomly chosen out of 31 million – we produce a CoNNL F1 score of 73.72 for entity detection on FIGER. Our end-to-end detection and typing evaluation produces macro and micro F1s of 45.11 and 54.75, based on the FIGER evaluation metrics. This work provides a foundation for the rapid development of new knowledge extraction pipelines.

Keywords: Question answering · Named entity recognition · Fine grained entity typing · Knowledge extraction

1 Introduction

It is common to encounter new knowledge extraction tasks for new product lines or projects [19]. New extractions are often needed in domains which are

© The Author(s) 2022
O. Corcho et al. (Eds.): EKAW 2022, LNAI 13514, pp. 51–65, 2022.
https://doi.org/10.1007/978-3-031-17105-5_4

either too new (e.g. carbon capture and sequestration) or too niche (e.g. material properties for engineering) to have relevant training data or hand-annotated labels. Creating new training data for such tasks is costly and difficult [17].

To tackle this problem, we propose using Question Answering (QA) as a strategy for low cost knowledge extraction with little to no additional training data. While numerous studies have explored the use of language models and question answering techniques for knowledge extraction, in most cases, these models are trained or fine-tuned on data specific to the new task [9,10,12].

In contrast, we start from the hypothesis that using pre-trained QA models with little to no additional training can effectively bootstrap domain specific entity extraction. We investigate this hypothesis, and explore how the addition of small amounts of training data could help lift model performance. This use of incremental addition of training data allows users to understand the trade-off between effectiveness of the model and the need to obtain more data.

Concretely, we start from a QA model trained on SQuAD 2.0 [15], and convert entity extraction and entity typing tasks into a QA format compatible with SQUAD for inference and for additional training. Importantly, to achieve this goal, we design and provide an open-source implementation of a framework for systematically applying QA to solve entity extraction tasks that deals in particular with both null and multiple answers.

To systematically evaluate the performance of QA models and the impact of additional training data for knowledge extraction, we use the task of fine-grained entity recognition and typing [11]. The aim of this task is to determine entity mentions and then assign them a type from a large set of potential predefined types. This task is appropriate as it provides a challenging proxy for real world environments where new long-tail entities need to be recognized.

The contributions of this paper are as follows:

- A framework that maps entity recognition tasks to question answering supporting BIO-type span tagging and that is able to use transformer-based QA models for the prediction of multiple answers per question that effectively deals with nulls. We address entity mention and type detection as an end-to-end problem, a very challenging task that is rarely covered in the literature.
- Measurement of the incremental gains achieved by small amounts of task-specific training data compared to a base SQuAD2.0 trained model in a fine-grained entity recognition setting.

This article is organized as follows. Section 2 describes related work. Section 3 introduces the datasets we employ. Section 4 continues with a discussion of our models, evaluation, and results. In Sects. 5 and 6 we provide a more detailed analysis of our results and reflect on the implications of our work.

2   Related Work

Information extraction in areas with little to no training data is a research area of growing importance [4]. Much work in this area focuses on distant or weak supervision. We test Question Answering for such low-resources situations, and use Fine-Grained Entity Recognition to evaluate our results. We discuss these two areas in-turn.

Question Answering: Question answering techniques are increasingly being used for information extraction. Perhaps the best known Question Answering dataset is SQuAD, the Stanford Question Answering Dataset [15]. SQuAD 1.1 consists of over 100,000 question answer pairs crowdsourced from hundreds of Wikipedia articles. These are typically used to train systems designed to extract information from text or perform other Natural Language Understanding and Reading Comprehension tasks. SQuAD asks questions about historical events, sports, geography, politics, and many other popular topics. Other datasets have followed, such as Discrete Reasoning Over Paragraphs (DROP) [5], which poses questions about sporting events that include numerical reasoning and comparison, and QAngeroo [20], which requires multi-Hop reasoning across multiple documents to assemble answers. He, et al. [7] reformulated Semantic Role Labeling as such a task. Levy, et al. [9] demonstrate the use of templated question answering for relation extraction. Most closely related to our work, Qi, et al. [12] and Li, et al. [10] show how multi-hop or multi-turn questions can allow machine comprehension models to resolve complex dependencies and compile multiple pieces of related information. We build on these ideas to tackle Fine-grained Entity Mention Detection and Type Detection as a single, end-to-end task.

Fine Grained Entity Mention Detection and Type Detection: Fine-grained Entity Mention Detection and Type Detection is a class of entity recognition task originating in Ling and Weld's 2012 Fine-Grained Entity Recognition (FIGER) paper, which observed that most Entity Recognition datasets were based on a very small number of entity types, plus a catch-all category of MISC [11]. Even some of the larger type vocabularies of the time, such as OntoNotes, only had a few dozen entity types [8]. FIGER addresses this by developing a type vocabulary of 112 much finer grained types, grouped into a two-level hierarchy.

FIGER includes hand-annotated gold data as well as distantly-supervised training data. Additionally, Ling and Weld develop a fine-grained entity recognition system. They report their systems performance on their gold data for end-to-end entity detection and typing, and also report the results of their model given the gold-data segmentation. Most subsequent research that is evaluated on the FIGER gold data only addresses the Fine-Grained Entity *Type Detection* task [2,18]. Additional work has expanded significantly on these type vocabularies, but has again focused only on the entity typing task [3]. Our work builds on the subset of research that uses the FIGER evaluation data to evaluate end-to-end entity *Mention Detection* and *Type Detection* pipelines. This includes Heuristics Allied with Distant Supervision (HAnDS), whose training data we build on [1]. More recently, Rodríguez, et al., also split the task into entity mention detection and type detection, but they treat them as distinct tasks and do not attempt an end-to-end solution [16].

3   QuAART Framework

Figure 1 illustrates our overall Question Answering with Additive Restrictive Training (QuAART) framework. Given a new type, the first step is to construct questions from templates based on the "type" of entity or property sought. Specifically, the question template generates questions in the form of "What was the [*type*]?" for each type in the vocabulary. The resulting questions are then fed to the question answering model with the associated passages of text. The answer to the question are a set of spans of text identifying the entity of the given type encoded in the question.

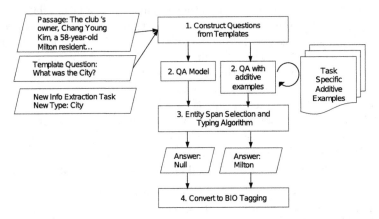

Fig. 1. The QuAART framework

A central component of the framework is knowing when *not* to answer the question, since we ask many questions for which we expect null results. Given the passage in Fig. 1, and the question "What was the spacecraft?", the question is unanswerable because there is no spacecraft mentioned in the passage. An unanswerable question returns a null result. With hundreds of types, QuAART poses hundreds of questions for each passage. It is critical to only produce answers where confidence is high.

To tackle this problem, we devised an algorithm to filter and select the most appropriate spans. The algorithm shown in Listing 1, uses heuristics to remove long answer spans (Line 7) that are likely not the names of entities and prefers entities that appear frequently (Line 13). Note, that this selection is done per text and per templated question.

An important input to the algorithm is a confidence threshold given to the model. QA models typically are not designed for entity detection tasks, so their model confidence thresholds for the predictions are set too low for this task. This results in many null answers for questions. To address this, we empirically determine an appropriate confidence threshold by using a small development set of labelled data. We note this confidence threshold does not necessarily need to be tuned for every new domain.

After entity recognition, the framework converts the results to the standard Begin-Inside-Outside (BIO) tagging system for evaluation. To this point, we have described the framework's use in a setting with a given QA model. However, the framework is also designed to enable the systematic retraining of datasets with task specific training data. Here, the key component is reformatting entity recognition datasets in a format that can be used to fine-tune QA models. We now describe the datasets used in our experiments based on this framework.

Algorithm 1: Entity span selection and typing

input : QAModel- A question answering model, that returns a set of answer spans given a passage of text, a question, and a confidence threshold;
Overlap- Given a set of answer spans,
find and return pairs of spans which overlap;
c - A confidence threshold;
D - Data in the form of a set of text passages;
T - A set of types to recognize

output: R - a map, $D \rightarrow \{(S,T)\}$, that maps each passage to an entity answer span and its associated type. S is the set of possible answer spans.

```
1  begin
2  |  R ← ∅ ;
3  |  for d ∈ D do
4  |  |  for t ∈ T do
5  |  |  |  q ← question template parameterized by t
6  |  |  |  A ← QAModel (d, q, c) ;
      |  |  |  // Remove long answer spans
7  |  |  |  for a ∈ A do
8  |  |  |  |  for x ∈ A \ a do
9  |  |  |  |  |  for y ∈ A \ a do
10 |  |  |  |  |  |  if |x| < |a| and |y| < |a|
11 |  |  |  |  |  |  and a is overlapping with x and y then
12 |  |  |  |  |  |  |  A ← A \ a

      |  |  |  // Pick a preferred overlapping span
13 |  |  |  foreach (x,y) ∈ Overlap(A) do
14 |  |  |  |  if freq(x) > freq(y) then A ← A \ y;
15 |  |  |  |  else if freq(y) > freq(x) then A ← A \ x;
16 |  |  |  |  else if freq(y) = freq(x) then
17 |  |  |  |  |  if |x| > |y| then A ← A \ y ;
18 |  |  |  |  |  else A ← A \ x;

19 |  |  |  foreach a ∈ A do
      |  |  |  |  // Update result with selected type and answer
20 |  |  |  |  R[d] ← +(a, t)
```

4   Datasets

Fine grained entity recognition tasks – especially when performed end to end – provide a challenging context for evaluating our framework. The evaluation data from FIGER data is among the most commonly used evaluation datasets in this research space [11]. The HAnDS dataset builds on FIGER, has evaluation data that uses similar types to FIGER, and, importantly, has a distantly-supervised training dataset that corresponds exactly to the type vocabulary in their evaluation data.

We provide a statistical description on FIGER and HAnDS below. Table 1 summarizes this information and Sect. 4 briefly describes the derivative datasets used in our experiments.

Table 1. Statistical summary of FIGER and HAnDS datasets

Dataset	Passage Count	Number of Entities	Distinct Types
FIGER Gold	434	563	43
HAnDS Gold	982	2,420	117
HAnDS Train	31,896,989	37,734,727	117

FIGER Data: The FIGER gold evaluation data consists of 434 sentences tagged with 563 entities using 43 entity types. FIGER also provides distantly-supervised training data generated from Wikipedia anchor texts [11]. This training dataset consists of two million passages. The mentions labeled in these passages use 8,566 distinct types, but not one of these passages limit mentions to the 113 official FIGER types. Given that QuAART only ask questions for, and can therefore only predict, in-vocabulary types, we do not use the FIGER training data. This is in-line with other approaches that use alternative training data and evaluate on FIGER [13].

HAnDS Data: HAnDS uses a type vocabulary of 118 types as opposed to FIGER's 113. The HAnDS types are not an exact superset of FIGER's: nine HAnDS classes are not present in FIGER, while four FIGER classes are not present in HAnDS.

The HAnDS evaluation data consists of 982 passages, split into a dev and test set of 446 and 536 passages respectively. The total evaluation dataset includes 2,420 entities tagged using 117 out of 118 types. The HAnDS training data is much larger than FIGER's, consists of 31 million passages, again from Wikipedia, but with entities tagged using the same 117 types as the evaluation data. Again, the training data is tagged using distant supervision.

Derived Question Answering Training Data: We construct a set of training data useful for fine tuning question answering for entity recognition. Specifically, we randomly select a tiny fraction – less than 0.0015% – of the HAnDS training data to build Question Answering data in a format that is compatible with SQuAD 2.0. This data is built in incremental chunks, adding 87 training *contexts/passages* at a time for 5 sets, totalling 435 passages. After compiling the first 5 sets, a 6th set was created adding another 34 passages. This additional set was to ensure that the final training data set included positive, answerable examples for all 118 types.

As per the QuAART framework, 118 questions are created, one per type in the HAnDS type vocabulary. The vast majority of the questions are not answerable and have null answers. Since SQuAD does not support multiple correct answers per question, this conversion is not lossless. In cases where there are more than one span of a given type in the source data, the resulting SQuAD-like will be missing some types and may even be missing entire entities. If there are two entities in the passage tagged with the */person* type, only one will be in the training data. Similarly, if there is a *person* entity co-occurring with another entity tagged */person* and */person/artist*, the second entity will only appear for the "Who was the artist?" question.

Table 1 below shows statistical distributions of the 6 training data files. There are always 118 questions per passage, but the vast majority of questions have null answers. The "Non-null questions" questions column counts the questions that have non-null answers. Similarly, non-null types counts the types that are effectively covered by non-null questions in the training set.

Table 2. Counts of HAnDS-specific Passages, Questions, "Possible" Questions, and "Possible" Types

Model	Passage count	Questions	Non-null answers	Non-null Types
SQuAD Only	0	0	0	0
SQuAD + 87	87	10266	159	51
SQuAD + 174	174	20532	320	67
SQuAD + 261	261	30798	517	77
SQuAD + 348	348	41064	725	83
SQuAD + 435	435	51330	889	84
SQuAD + 468	468	55342	1045	117

5   Experimental Method and Results

We run two sets of experiments. Data source information, data conversion scripts, and evaluation scripts as well as information on model training and inference can be found on the QuAART GitHub Repository.[1] In *Experiment 1*, we fine-tune against HAnDS training data and evaluate against both the FIGER and HAnDS evaluation sets.

The HAnDS training data is distantly supervised. In production settings, small amounts of gold labeled data may be more available than large corpora of distantly supervised data. Therefore, it is important to understand the impact of using hand labeled gold data for training. In *Experiment 2*, we construct

---

[1] https://github.com/elsevierlabs-os/quaart.

train/dev/test splits out of the existing hand annotated FIGER evaluation data. For both experiments, we report two sets of scores:

1. Entity Mention Detection scores - this determines how well the model performs in detecting mentions of entities in text ignoring types. Specifically, we use the Conference on Natural Language Learning (CoNNL) F1 metric treating every entity as type MISC.
2. Entity Type Detection scores - this is the end-to-end performance on the task of recognizing entity mention and assigning an appropriate type. Here, we report FIGER's Strict, Loose Macro F1, and Loose Micro F1 scores, as implemented in Shimaoka, et al. [11,18].

5.1   Experiment 1: Incremental Training with HAnDS

A RoBERTa model fine tuned on SQuAD 2.0 is taken as a base. Progressively larger sets of HAnDS training data are added and the model is fine-tuned from the base for each increment of data. Given the length of training, we only perform one sampling. We provide our splits in the GitHub repository. After each model retraining, predictions are run against dev splits of both HAnDS and FIGER.

At training time, we use max sequence length increased to 512 to support the longer passages found in both the HAnDS and FIGER datasets. Inference also uses a max_seq_length of 512, and an n_best of 10 to slightly constrain the possible sets of answers produced.

As noted in Sect. 3, the HAnDS evaluation data already comes split into dev and test sets. This is *not* the case for FIGER, so a dev split is generated containing slightly more than 10% of the overall evaluation data. For each of the models above, predictions are run against the FIGER and HAnDS dev splits. As discussed in Sect. 3, these dev sets are used to tune post-processing routines and heuristics for generating BIO tagged sequences from the SQuAD Question Answering Results.

Specifically, the standard SQuAD predictions do not fit the use case of fine-grained entity mention detection and type detection, as they assume one answer per passage. Additionally, model confidence thresholds for the predictions are far too low, resulting in almost entirely null answers. For vanilla SQuAD, these thresholds are slightly higher, but they drop significantly after being exposed to thousands of additional null answer examples from the HAnDS training data. Instead of using the predictions as is, we process the n_best prediction sets. This allows for a tuneable prediction threshold that can vary from model to model. More significantly, this provides a mechanism for potentially generating more than one answer per question in cases where multiple entities of the same type exist in one passage. As noted previously, this multiple-entity scenario is common in both the HAnDS and FIGER datasets.

The confidence threshold with the best performance for each model on the dev sets is used when running predictions for the full evaluation sets for both FIGER and HAnDS.

Results: Tables 3 and 4 give the results for both evaluation datasets on both the Entity Mention Detection task, and the Entity and Type Detection task. As a reminder these results are using the HAnDS training data.

Table 3. Entity Mention Detection F1 scores for both FIGER and HAnDS.

Model	F1 FIGER	F1 HAnDS
SQuAD only	0.37	0.47
SQuAD + 87	0.70	0.54
SQuAD + 174	0.72	0.58
SQuAD + 261	0.66	0.59
SQuAD + 348	0.70	0.59
SQuAD + 435	0.74	0.62
SQuAD + 468	0.70	0.63

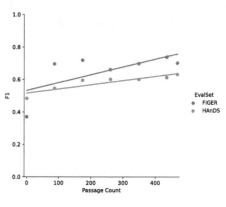

Fig. 2. Mention Detection scores steadily increase with additional training data.

For Entity Mention Detection evaluated on FIGER, the initial increment of training data nearly doubles the scores achieved by the QA model trained on SQuAD alone. Subsequent additions of data offer less improvement, and in some cases lower performance. In the HAnDS dataset, though the initial boost is smaller, the results do continue to rise with each progressive addition of training data. The same trends hold true for the end-to-end detection plus typing results in Table 4.

Table 4. End-to-end Detection and Typing scores on FIGER and HAnDS.

Model	FIGER Evaluation			HAnDS Evalution		
	Strict F1	Micro F1	Macro F1	Strict F1	Micro F1	Macro F1
SQuAD Only	0.04	0.18	0.11	0.05	0.26	0.18
SQuAD + 87	0.27	0.51	0.42	0.09	0.31	0.22
SQuAD + 174	0.30	0.55	0.45	0.11	0.36	0.26
SQuAD + 261	0.27	0.50	0.39	0.11	0.36	0.26
SQuAD + 348	0.24	0.52	0.41	0.13	0.39	0.27
SQuAD + 435	0.27	0.52	0.44	0.11	0.37	0.26
SQuAD + 468	0.19	0.48	0.39	0.13	0.42	0.30

To further understand the impact of incremental data, Figs. 2, 3a, 3b fit a linear regression to distributions for CoNNL, FIGER Micro and FIGER Macro F1 Scores. On all three metrics, adding the first iteration of HAnDS training data creates substantive increases in score. This is especially pronounced for the FIGER evaluation scores, which largely level out and even decrease slightly for some of the training data additions.

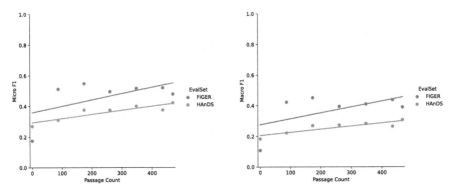

(a) FIGER Micro F1 scores generally increase with additional training data.

(b) FIGER Macro F1 scores generally increase with additional training data.

Fig. 3. FIGER regressions

These results clearly demonstrate that just using SQuAD provides a reasonable and very low effort starting point for entity detection and fine-grained entity typing in new domains. Only a list of types needs to be prepared. With even a small addition of training data the quality of information extraction improves. Further additions of training data, while still useful, only marginally improve results.

We limited our experiments to the addition of up to 500 passages due to the increasing run-time required for training models. Our largest model takes two to three hours to train. Since each passage results in an addition of one question *per type*, training data can have tens of thousands of questions for only hundreds of passages.

Table 5 shows our best performing models (SQuAD + 174 for FIGER and SQuAD + 468 for HAnDS) compared against the original FIGER and HAnDS papers, which are the only other two studies we are aware of that attempt perform end-to-end Entity Mention Detection and Entity Type Detection on these datasets. We refer to the FIGER model from Ling and Weld [11] as Distant Supervision (DS), and the HAnDS model from Abhishek, et al. [1] as Distant Supervision with Heuristics (DSH). DS uses 2 million training examples and DSH uses 31 million examples to achieve their results. The QuAART approach uses a fraction of that data: 174 passages (0.0005%) of HAnDS training data in our top performing FIGER model, and 468 (0.0015%) when evaluating on HAnDS.

Table 5. End-to-end Detection and Typing scores situated against other systems.

FIGER Evalution Data					HAnDS Evalution Data			
Model	Strict F1	Macro F1	Micro F1		Model	Strict F1	Macro F1	Micro F1
DS [11]	0.47	0.62	0.60		DSH [1]	0.53	0.68	0.69
DSH [1]	0.56	0.71	0.68		SQuAD + 468	0.13	0.42	0.31
SQuAD + 174	0.30	0.55	0.45					

5.2 Experiment 2 – Training with Gold Data

Given that it took remarkably few passages to start seeing viable results in Experiment 1, we wanted to investigate the use of gold training data that was not produced using distant supervision. Experiment 2 was conducted using the FIGER evaluation dataset, which we further subdivided into separate training, development, and test sets. We kept the bulk of the data in test (326 of the 434 total passages), and used dev and training sets of 54 passages each. The training test was further subdivided into 9 random batches of 6 passages each. This splitting was done 3 different times as to limit the effect of specific training examples on the data ablations.

Results: Figure 4 shows our Entity Mention Detection F1, and end-to-end Mention & Typing Macro and Micro F1 scores through all of these shuffles. Similar to above, the incremental addition of small amounts of training data improve performance. It is noteworthy that some of the higher scores come with very little training data. The top mention detection scores are from models trained with only 18 passages of text. As passages are added, mention detection scores drop slightly, while end-to-end mention and type detection scores gain slightly.

6 Discussion

QuAART discusses *restrictive* examples because much of the benefit of the added data is in reigning in false positives on the end-to-end task. For example, the SQuAD only model might predict /person, /person/actor, and /person/musician for an entity with a gold type of only /person. The presence of large numbers of null answers for these rarer types, in the additive examples, reduces the likelihood of these false positives.

Beyond the original SQuAD v.2 paper, which introduced the null answers, little has been written about the value of this null response [14]. To our knowledge, no further investigations have been done into how the generation of null response questions can improve the results of other information extraction tasks.

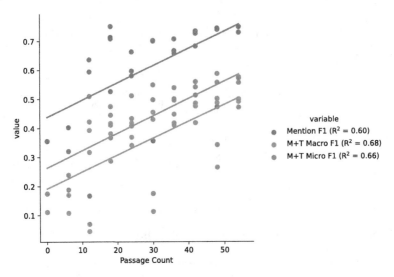

Fig. 4. Cross-validated FIGER scores trained on small amounts of gold data. F1 scores are entity mention detection scores, while Micro and Macro F1 are for end to end entity mention detection and type detection (M+T).

As shown in Table 2, the overwhelming majority of our additive examples are negative examples. On average, only around 1.5% of the questions generated from the HAnDS data have an answer. This tends to reduce the overall number of total predictions, increasing precision, though slightly reducing recall.

While our fine-tuned models achieve much higher scores overall, they also predict fewer classes. On FIGER, our SQuAD only model, predicts 1,090 spans across 94 classes, but only matches 164 spans correctly. SQuAD + 174 predicts 599 spans across only 28 classes, but matches 367 of those spans correctly.

Table 6 isolates scores for a few types of "organization" before and after adding 174 HAnDS passages. For this set of examples, the F1 scores go up for *every* class, regardless of whether the number of predicted spans increase or decrease. Both precision and recall improve substantively.

Table 7 looks more closely at a specific example. The SQuAD only model misses more entities, predicts an erroneous entity, and specifically overpredicts types. The addition of a mere 174 passages of HAnDS examples results in predictions that are much closer to the gold data, and the errors produced by the model – such as location/city for Utah – make much more intuitive sense.

These results show promise for future work. Specifically, we aim to investigate whether using the relationship between fine grained types and more generic types can improve performance. Additionally, there is scope to applying this approach to extract other important knowledge such as relations or attributes [6].

Table 6. Scores for "organization" types before and after adding 174 HAnDS passages (Preds is count of predictions, Matches is count of matching predictions)

Model	Type	Preds	Matches	Strict F1	Micro F1	Macro F1
SQuAD Only	Organization	69	25	0.05	0.14	0.11
SQuAD+174	Organization	123	65	0.14	0.38	0.28
SQuAD Only	Company	25	6	0.00	0.10	0.09
SQuAD + 174	Company	11	5	0.09	0.22	0.14
SQuAD Only	Ed. Institution	24	6	0.02	0.14	0.11
SQuAD + 174	Ed. Institution	5	3	0.04	0.17	0.10
SQuAD Only	Sports League	5	1	0.00	0.04	0.06
SQuAD + 174	Sports League	3	3	0.00	0.32	0.22

Table 7. Example passage, gold data, and predictions from FIGER eval dataset.

Passage: The biggest cause for concern for McGuff is the bruised hamstring Regina Rogers suffered against Utah last Saturday .	
Gold	**McGuff**: /person **Regina Rogers**: /person, /person/athlete **Utah**: /organization,/organization/sports_team **Saturday**: /time
SQuAD Only	**bruised hamstring**: /product,/event/attack, /medical_treatment,/symptom **Regina Rogers**: /person,/person/actor, /person/artist,/person/athlete,/person/soldier **Utah**: /product/game
SQuAD + 174	**McGuff**: /person **Regina Rogers**: /person, /person/athlete **Utah**: /organization,/location,/location/city,/time

7   Conclusion

We present QuAART, a framework for mapping entity recognition tasks to question answering tasks. QuAART includes the construction of questions from templates, an algorithm for selecting high-confidence answers, and a system for mapping back to BIO tags for evaluation. The framework is used to test the performance of question answering models for the task fine-grained entity mention and type detection. We start from a model trained on SQuAD 2.0, and iteratively add small amounts of training data from HAnDS, tracking the improvements achieved through each iteration. We run a second experiment using a small training split of the hand-labeled FIGER evaluation data, which more closely approximates real-world information extraction tasks.

Our results show that question answering can be a viable approach for quickly constructing new knowledge extraction pipelines. Users need only formulate a list of entity types and generate questions in order to extract new information. This is faster and simpler than labelling data or constructing large distantly supervised corpora. Importantly, we show that with only a small amount of domain specific question answering training data performance can be improved allowing users to find a balance between quick construction of a pipeline and extraction performance.

Acknowledgments. The authors would like to thank Curt Kohler and Antony Scerri for various discussions and reviews of this work. This project was funded in-part by Elsevier's Discovery Lab.

References

1. Abhishek, A., Taneja, S.B., Malik, G., Anand, A., Awekar, A.: Fine-grained entity recognition with reduced false negatives and large type coverage. In: Automated Knowledge Base Construction (AKBC) (2019)
2. Chen, Y., et al.: An empirical study on multiple information sources for zero-shot fine-grained entity typing. In: Proceedings of the 2021 Conference on Empirical Methods in Natural Language Processing (EMNLP 2021), pp. 2668–2678, November 2021
3. Choi, E., Levy, O., Choi, Y., Zettlemoyer, L.: Ultra-fine entity typing. In: Proceedings of the 56th Annual Meeting of the Association for Computational Linguistics (ACL 2018), pp. 87–96, July 2018. https://doi.org/10.18653/v1/P18-1009
4. Deng, S., Zhang, N., Chen, H., Xiong, F., Pan, J.Z., Chen, H.: Knowledge extraction in low-resource scenarios: Survey and perspective (2022). https://arxiv.org/abs/2202.08063
5. Dua, D., Wang, Y., Dasigi, P., Stanovsky, G., Singh, S., Gardner, M.: DROP: a reading comprehension benchmark requiring discrete reasoning over paragraphs. In: Proceedings of the 2019 Conference of the North American Chapter of the Association for Computational Linguistics, pp. 2368–2378, June 2019
6. Harper, C., Cox, J., Kohler, C., Scerri, A., Daniel Jr., R., Groth, P.: SemEval-2021 task 8: MeasEval - extracting counts and measurements and their related contexts. In: Proceedings of the 15th International Workshop on Semantic Evaluation (SemEval-2021), pp. 306–316, August 2021
7. He, L., Lewis, M., Zettlemoyer, L.: Question-answer driven semantic role labeling: Using natural language to annotate natural language. In: Proceedings of the 2015 Conference on Empirical Methods in Natural Language Processing, pp. 643–653, September 2015
8. Hovy, E., Marcus, M., Palmer, M., Ramshaw, L., Weischedel, R.: Ontonotes: the 90% solution. In: Proceedings of the Human Language Technology Conference of the NAACL. NAACL-Short 2006, USA, pp. 57–60 (2006)
9. Levy, O., Seo, M., Choi, E., Zettlemoyer, L.: Zero-shot relation extraction via reading comprehension. In: Proceedings of the 21st Conference on Computational Natural Language Learning (CoNLL 2017), pp. 333–342, August 2017
10. Li, X., et al.: Entity-relation extraction as multi-turn question answering. In: Proceedings of the 57th Annual Meeting of the Association for Computational Linguistics, pp. 1340–1350, July 2019

11. Ling, X., Weld, D.S.: Fine-grained entity recognition. In: Proceedings of the Twenty-Sixth AAAI Conference on Artificial Intelligence, AAAI 2012, pp. 94–100. AAAI Press (2012)
12. Qi, P., Lin, X., Mehr, L., Wang, Z., Manning, C.D.: Answering complex open-domain questions through iterative query generation. In: Proceedings of the 2019 Conference on Empirical Methods in Natural Language (EMNLP-IJCNLP 2019), pp. 2590–2602, November 2019
13. Qian, J., et al.: Fine-grained entity typing without knowledge base. In: Proceedings of the 2021 Conference on Empirical Methods in Natural Language Processing (EMNLP 2021), Online and Punta Cana, Dominican Republic, pp. 5309–5319, November 2021
14. Rajpurkar, P., Jia, R., Liang, P.: Know what you don't know: unanswerable questions for SQuAD. In: Proceedings of the 56th Annual Meeting of the Association for Computational Linguistics, pp. 784–789, July 2018
15. Rajpurkar, P., Zhang, J., Lopyrev, K., Liang, P.: SQuAD: 100,000+ questions for machine comprehension of text. In: Proceedings of the 2016 Conference on Empirical Methods in Natural Language Processing, November 2016
16. Rodríguez, A.J.C., Castro, D.C., García, S.H.: Noun-based attention mechanism for fine-grained named entity recognition. Expert Syst. Appl. **193** (2022). https://doi.org/10.1016/j.eswa.2021.116406
17. Roh, Y., Heo, G., Whang, S.E.: A survey on data collection for machine learning: a big data-AI integration perspective. IEEE Trans. Knowl. Data Eng. **33**, 1328–1347 (2021). https://doi.org/10.1109/TKDE.2019.2946162
18. Shimaoka, S., Stenetorp, P., Inui, K., Riedel, S.: Neural architectures for fine-grained entity type classification. In: Proceedings of the 15th Conference of the European Chapter of the Association for Computational Linguistics, pp. 1271–1280, April 2017. https://aclanthology.org/E17-1119
19. Surdeanu, M., McClosky, D., Smith, M., Gusev, A., Manning, C.: Customizing an information extraction system to a new domain. In: Proceedings of the ACL 2011 Workshop on Relational Models of Semantics, pp. 2–10, June 2011
20. Welbl, J., Stenetorp, P., Riedel, S.: Constructing datasets for multi-hop reading comprehension across documents. Trans. Assoc. Comput. Linguist. **6**, 287–302 (2018)

New Strategies for Learning Knowledge Graph Embeddings: The Recommendation Case

Nicolas Hubert[1,2]([✉]) [ID], Pierre Monnin[3] [ID], Armelle Brun[1] [ID], and Davy Monticolo[2] [ID]

[1] Université de Lorraine, CNRS, LORIA, Nancy, France
{nicolas.hubert,armelle.brun}@loria.fr
[2] Université de Lorraine, ERPI, Nancy, France
{nicolas.hubert,davy.monticolo}@univ-lorraine.fr
[3] Orange, Belfort, France
pierre.monnin@orange.com

Abstract. Knowledge graph embedding models encode elements of a graph into a low-dimensional space that supports several downstream tasks. This work is concerned with the recommendation task, which we approach as a link prediction task on a single target relation performed in the embedding space. Training an embedding model requires negative sampling, which consists in corrupting the head or the tail of positive triples to generate negative ones. Although knowledge graph embedding models and negative sampling have extensively been investigated for link prediction, their combined use for performing recommendations over knowledge graphs remains largely unexplored in the literature. In this work, we propose two specialization strategies for training embedding models and performing knowledge graph-based recommendations. Both strategies first train an embedding model on the whole knowledge graph. Then, during a specialization phase, a dedicated negative sampling scheme is applied to refine the pre-trained model. Experimental results on two public datasets demonstrate that a simple strategy which refines a pre-trained model by sampling random negative tails for the target relation proves to be very effective. This strategy significantly improves performance with respect to traditional rank-based evaluation metrics as well as a newly introduced metric that reflects the semantic validity of the top-ranked candidate entities.

Keywords: Negative Sampling · Knowledge Graph Embedding · Recommendation · Link Prediction · Ontology

1 Introduction

A knowledge graph (KG) is a collection of facts (h, r, t) where h (head) and t (tail) are two entities of the graph, and r is the semantic relation that links them. KGs are used for several tasks including entity matching, question answering

© The Author(s), under exclusive license to Springer Nature Switzerland AG 2022
O. Corcho et al. (Eds.): EKAW 2022, LNAI 13514, pp. 66–80, 2022.
https://doi.org/10.1007/978-3-031-17105-5_5

and link prediction [24]. The latter is the focus of this paper. Link prediction consists in assessing the probability of existence of a given triple, for example in a knowledge graph completion perspective. Several approaches address the link prediction task, especially Knowledge Graph Embedding (KGE) methods [18,22]. They encode entities and relations of the KG into a low-dimensional vector space that preserves the structure of the original graph [22].

Training such KGE models requires both positive and negative triples. As KGs are usually made up of only positive triples, negative sampling (NS) is used to generate non-existent triples by corrupting the head or the tail of positive triples with any other entity from the KG [18]. Resulting triples are called negative samples and they constitute the basis on which embedding learning is performed: embedding models iteratively learn to assign higher ranks to true triples than to negative ones. Hence, the way these models learn is significantly influenced by negative sampling methods, which therefore received much attention recently [7,8,12].

These negative sampling methods usually intervene in link prediction tasks that consider all relations in the KG. However, in some application domains there is a target relation, i.e. a certain type of link that is of interest for prediction. In this work, we address the recommendation task and we approach it as a link prediction task on a single relation. More specifically, the target relation represents the link between users and items to recommend [4,17]. Recommendation consists in predicting one or a few relevant items of interest for a current user. In other words, for a given triple $(u, r, ?)$, the goal is to recommend the more relevant items for user u, with r denoting the nature of the recommendation based on the application context. For instance, a use case could be recommending university curricula to high school students. Although KGE models have successfully been used for recommendation [3,17], literature has primarily focused on KGE and NS for the more generic link prediction task. Predicting on a unique relation arguably responds to a different learning objective. Thus, we claim that new approaches are needed and we formulate the following research question:

RQ1. Does refining a KGE model pre-trained on a generic link prediction task by specializing training on the target relation improve recommendation performance?

Once a KGE model is trained, its quality with regard to the link prediction task is traditionally assessed using rank-based metrics. These metrics face some limitations as they only focus on the presence (or not) of the ground-truth in the top-K list. From our view, the ability of a model at predicting links that are semantically close to the ground-truth is a supplementary dimension that needs to be evaluated in order to have a more comprehensive view of a model quality. This way, the ability of a model to retain the semantic profile (i.e. range type) of the target relation can be better assessed. Please note that in the rest of the paper, we only consider the range type of the relation when referring to its semantic profile. This raises our second research question:

RQ2. What is the impact of different training strategies on the ability of KGE models to capture the semantic profile of the target relation?

Regarding **RQ1**, our motivation is to study one-hop KGE models that proved successful in downstream recommendation tasks [3, 17], and determine whether their performance can be enhanced by specializing the training procedure. **RQ2** goes a step further, and we ask whether a more informed negative sampling method has a positive impact on the ability of an embedding model to retain the semantic profile of the target relation. The main contributions of our work are summarized as follows.

- We introduce two novel strategies for training knowledge graph embeddings for a downstream recommendation task.
- We introduce a new metric that measures the semantic validity of the top-ranked candidate entities.

The remainder of the paper is structured as follows. Related work about KGE models for recommendation and negative sampling is presented in Sect. 2. In Sect. 3, we detail the models and the proposed strategies for making KGE-based recommendations. Dataset descriptions, experimental settings and key findings are provided in Sect. 4. A discussion is provided in Sect. 5. Lastly, Sect. 6 summarizes the key findings and outlines directions for future research.

2   Related Work

2.1   Knowledge Graph Embeddings for Recommendation

Several methods have been proposed for making recommendations over KGs. Most recent approaches based on Graph Neural Networks (GNNs) showcase impressive performance [27]. Their success in recommendation tasks derives from their ability to model higher-order connectivity and encode sparse semi-supervised signals [5]. However, GNN-based recommender systems have some limitations: due to their large memory usage and significant training time, GNNs are not always applicable in real-world scenarios [27].

On the contrary, one-hop embedding models are simpler models that proved to be successful in downstream recommendation tasks [15]. Grad-Gyenge *et al.* [6] compare traditional collaborative filtering and embedding models for making recommendations over KGs. They clearly show that the latter significantly increase recommendation performance without suffering from an increasing amount of user interactions, contrary to traditional collaborative filtering algorithms. However, their experiments do not consider mainstream KGE models that are most commonly used. By contrast, [17] analyses the experimental results obtained with popular KGE models. Their work is close to our line of research, as it is concerned with the recommendation task. Although the authors clearly emphasize the superiority of KGE models over traditional baselines in a recommendation framework, they only focus on translational models which include TransE [1], TransH [25] and TransR [14], and do not consider other popular and effective KGE models [18]. In addition, they do not study the influence of negative sampling on recommendation performance.

2.2   Negative Sampling for Link Prediction in Knowledge Graphs

While KGs usually comprise positive triples only, training KGE models requires negative samples [12]. An early introduced method is Random Negative Sampling (RNS). It consists in replacing either the head h or the tail t of a triple with any other entity sampled uniformly from the set of observed entities in the KG [1]. However, sampling random entities uniformly can generate positive triples. For example, replacing the tail of (CristianoRonaldo, playedFor, RealMadrid) with ManchesterUnited would generate the triple (CristianoRonaldo, played-For, ManchesterUnited) which actually represents a true fact. The approach presented in [25] reduces the risk of creating such false-negative triples by setting different probabilities for replacing the head and the tail based on the nature of the relation that links them: if the relation r is 1-to-N (e.g. parentOf), the head h has a higher probability of being replaced. If the relation r is N-to-1 (e.g. bornIn), the tail t is more likely to be replaced. More specifically, [25] uses a Bernoulli distribution to sample heads or tails with distinct probabilities.

However, the two aforementioned approaches cannot prevent the sampling procedure from producing semantically incorrect triples such as (Cristiano-Ronaldo, playedFor, SoccerShoes). Such nonsensical triples do not provide the model with sufficient signal to learn from, which causes the notorious zero loss problem [23]. Therefore, more sophisticated methods have been proposed to generate high-quality negative samples and consequently give more hints to the model training [30]. In particular, ontological constraints and domain knowledge can be leveraged to create meaningful negative samples [10,13,26]. Intuitively, generating more realistic and robust negative samples helps the embedding model learn a better vector representation of the graph components. For instance, type-constrained negative sampling (TCNS) [13] replaces the head or the tail with a random entity belonging to the same type as the ground-truth entity. By doing so, only semantically valid triples are generated during negative sampling. TCNS has been found to work better than pure RNS in several scenarios [12,13]. However, it should be noted that entities are rarely typed [2,13]. To the best of our knowledge, RNS and TCNS have not been studied in the specific frame of single-relation link prediction, especially recommendation. It would be interesting to investigate whether they demonstrate greater efficiency in such a framework.

3   Training Embedding Models for Recommendation

Figure 1 outlines the whole approach that we further explicit in the following. In Sect. 3.1, we summarize the KGE models used in this work. Then, in Sect. 3.2 we elaborate on the specialization training and negative sampling strategies that we propose. These strategies are designed to fit the recommendation task. Finally, Sect. 3.3 details a newly introduced semantic-oriented metric that – combined with traditional rank-based metrics such as Hits@K – provides a more comprehensive view of the quality of a KGE model.

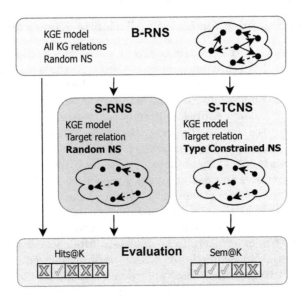

Fig. 1. Approach outline. S-RNS and S-TCNS are the two specialization strategies that we propose in this work. Both only include the target relation in the training procedure (dashed arrows), but differ regarding the negative sampling method. These two strategies are benchmarked against B-RNS in terms of Hits@K and our newly introduced semantic-oriented metric Sem@K.

3.1   Knowledge Graph Embedding Models

In line with [12], we use TransE [1], TransH [25], DistMult [28], and Com-plEx [21], which all are very popular models that have proven to work well in a wide range of link prediction tasks. These models are detailed below.

TransE is the earliest translational model. It learns representations of enti-ties and relations such that for a triple (h, r, t), $\mathbf{e}_h + \mathbf{e}_r \approx \mathbf{e}_t$ where $\mathbf{e}_h$, $\mathbf{e}_r$ and $\mathbf{e}_t$ are the head, relation and tail embeddings, respectively. The scoring function is $f(h, r, t) = -d(\mathbf{e}_h + \mathbf{e}_r - \mathbf{e}_t)$ with d a distance function, usually the $L1$ or $L2$ norm. TransE does not properly handle 1-to-N, N-to-1 nor N-to-N relations [24] and yet has been found to be very efficient in multi-relational settings [3].

TransH is an extension of TransE. It allows entities to have distinct repre-sentations when involved in different relations. Specifically, $\mathbf{e}_h$ and $\mathbf{e}_t$ are pro-jected into relation-specific hyperplanes with projection matrices $\mathbf{w}_r$. If (h, r, t) holds, the projected entities $\mathbf{e}_{h_\perp} = \mathbf{e}_h - \mathbf{w}_r^T \mathbf{e}_h \mathbf{w}_r$ and $\mathbf{e}_{t_\perp} = \mathbf{e}_t - \mathbf{w}_r^T \mathbf{e}_t \mathbf{w}_r$ are expected to be linked by the relation-specific translation vector $\mathbf{d}_r$. Thus, the scoring function is $f(h, r, t) = -d(\mathbf{e}_{h_\perp} + \mathbf{d}_r - \mathbf{e}_{t_\perp})$. TransH often showcases better performance than TransE with only slightly more parameters [25].

DistMult is a semantic matching model. It is characterized as such because it uses a similarity-based scoring function and matches the latent semantics of entities and relations by leveraging their vector space representations. More specifically, DistMult is a bilinear diagonal model that uses a trilinear dot product as its scoring function: $f(h, r, t) = \langle \mathbf{e}_h, \mathbf{W}_r, \mathbf{e}_t \rangle$. It is similar to RESCAL [16] – the very first semantic matching model – but restricts relation matrices $\mathbf{W}_r \in \mathbb{R}^{d \times d}$ to be diagonal. As the scoring function of DistMult is commutative, all relations are considered symmetric. This assumption does not hold in general. However, DistMult still achieves state-of-the-art performance in most cases [11].

ComplEx is also a semantic matching model. It extends DistMult by using complex-valued vectors to represent entities and relations: $\mathbf{e}_h, \mathbf{e}_r, \mathbf{e}_t \in \mathbb{C}^d$. As a result, ComplEx is better able to model antisymmetric relations than DistMult [19]. Its scoring function uses the Hadamard product: $f(h, r, t) = \mathrm{Re}\,(\mathbf{e}_h \odot \mathbf{e}_r \odot \bar{\mathbf{e}}_t)$ where $\bar{\mathbf{e}}_t$ denotes the conjugate of $\mathbf{e}_t$.

3.2   New Specialization Strategies for Training KGE

To train the aforementioned KGE models, we consider the strategies depicted in Fig. 1. Contrary to the generic link prediction task, we assume that for a downstream recommendation task, the training phase should focus more on the target relation. However, the KGE model still needs to be trained considering all entities and relations before specializing in order to take into account all the available information in the KG. This motivates the pre-training step included in the two proposed strategies: both of them reuse a generic KGE model that was trained on the whole graph. It means that the training phase is first performed on all training triples, regardless of their relation. In the experiments, the generic KGE model trained on the whole graph with traditional RNS [1] (top of Fig. 1) serves as our baseline and is referred to as **B-RNS** (standing for baseline random negative sampling).

Once the KGE model is pre-trained, triples whose relation differs from the target relation are filtered out from the train set. For both strategies presented below, training resumes by specializing the training of the KGE model on the resulting filtered train set. The way the specialization phase is achieved depends on the negative sampling scheme used:

S-RNS stands for specialized training with random negative sampling. In this strategy, the embedding model is refined by applying uniform RNS when corrupting the tails of the remaining train triples.

S-TCNS stands for type-constrained negative sampling. In this strategy, the embedding model is refined by applying TCNS [13] when corrupting the tails of the remaining train triples.

3.3   Evaluating Recommendations: Hits@K and Sem@K

Hits@K is a rank-based metric extensively used for link prediction tasks. This metric accounts for the proportion of positive triples that are ranked in the top-K positions against a set of negative triples. Hits@K is also used in recommendation [29]. As this work focuses on the recommendation task, the generic definition of Hits@K [18] has to be modified to account for the fact that only tails are corrupted. The modified Hits@K is defined in Eq. (1).

$$\text{Hits@}K = \frac{1}{|\mathcal{B}|} \sum_{(h,r,t)\in\mathcal{B}} \mathbb{1}\left(\text{rank}_{(h,r)}(t) \leq K\right) \tag{1}$$

where $\mathcal{B}$ is a batch of positive triples and $\text{rank}_{(h,r)}(t)$ denotes the position of the ground-truth tail t in the sorted list of top-K entities scored by a given KGE model, for the head h, and the target relation r.

However, from our view, Hits@K remains limited when considering the recommendation task. Indeed, it only represents the ability of a model to rank the ground-truth higher in the top-K list. As such, this metric does not fully answer the following questions:

- When two KGE models have similar Hits@K, can we refine the evaluation process to safely favour one model?
- When the ground-truth does not show up in the top-K list, how do we assess the extent to which the KGE model has captured the semantic profile of the target relation?

Traditionally, the aforementioned questions have been addressed by considering additional rank-based metrics such as Mean Rank (MR) and Mean Reciprocal Rank (MRR) [18]. Compared to Hits@K, they take into account the position of the ground-truth without a threshold K. However, all rank-based metrics solely focus on the ground-truth at hand, without taking into consideration the other ranked entities. In some application domains, knowing the rank of the ground-truth is not sufficient. In the recommendation case, the catalog of items can be huge, so that recommending entities that are related to the ground-truth is also of interest. In the context of a KG, similarity between entities can be reflected through their types. As a result, it is desirable for a KGE model to also retain the semantic profile of the target relation in order to assign higher scores to triples that are semantically close to the ground-truth.

To this aim, we introduce Sem@K, a new semantic-oriented metric. Combining it with Hits@K gives a more comprehensive view of the quality of a KGE model. Sem@K reflects to what extent a K-list contains semantically valid candidates with regard to the range of the target relation. Hence the definition of Sem@K in Eq. (2):

$$\text{Sem@}K = \frac{1}{|\mathcal{B}|} \sum_{q\in\mathcal{B}} \frac{1}{K} \sum_{q'\in\mathcal{S}_q^K} \text{compatibility}(q,q') \tag{2}$$

where $\mathcal{S}_q^K$ is the top-K list of candidate triples scored by a given KGE model given a ground-truth triple q, and $\text{compatibility}(q,q')$ (Eq. (3)) assesses whether the candidate triple q' is semantically compatible with its ground-truth counterpart q. As this work focuses on the recommendation task, by semantic compatibility we refer to the fact that the predicted tail belongs to the range of the target relation:

$$\text{compatibility}(q, q') = \begin{cases} 1, & \text{if type}(q'_t) = \text{range}(q_r) \\ 0, & \text{otherwise} \end{cases} \tag{3}$$

where type(e) returns the type of entity e and range(r) is the range of the relation r. q_r and q'_t denote the ground-truth relation and the tail of the candidate triple, respectively.

Compared to Hits@K that naturally increases or remains equal with higher values of K, Sem@K behaviour is non-monotonic. However, when the set of semantically valid candidates is limited, one should reasonably expect a decreasing Sem@K with higher K. Sem@K can be generalized to any other semantic context by adapting the semantic compatibility compatibility() operator with the context-dependent one. In our experiments, all strategies are evaluated using Hits@K and Sem@K, as depicted in Fig. 1.

4   Experiments

To address the relevance of the proposed specialization training strategies, experiments are conducted on real-world and public datasets and we subsequently evaluate these strategies in terms of Hits@K and the newly introduced Sem@K.

4.1   Datasets

As aforementioned, our contributions are evaluated on EduKG and KG20C, two KGs that we choose for their adequate entity typing: given a head-relation pair (h, r), the missing tail can only be of one single type. Consequently, we are able to study the influence of TCNS [13]. Both datasets naturally lend themselves to the recommendation task, whereas in most datasets used for link prediction, it is more questionable to favour one relation over the others. Below, we provide a thorough description of the KGs used in the experiments and indicate their respective target relation. Their main characteristics are summarized in Table 1.

EduKG[1] is an educational KG that authors of this work built and that is instantiated with students and a broad spectrum of university curricula. The goal is to recommend university curricula to high school students. More formally, for each triple of the form $(h, r, ?)$, we aim at retrieving the ground-truth tail t where h is an entity of type Student, r accounts for the relation likedCurriculum and t is an entity of type Curriculum. An example would consist in retrieving the correct tail for the following test triple: (Bob, likedCurriculum, ?). Hence, for the S-TCNS strategy, only entities of type Curriculum are used to replace tails of positive triples during negative sampling. In EduKG, both the users (students) and items (curricula) are highly connected to other entities of the KG. For instance, students are linked to a HighSchoolMajor, they are characterized by PersonalityTraits, they have a certain amount of favorite SchoolSubjects, and they provide Keywords reflecting their interests. Likewise, curricula are

---

[1] purl.org/edukg/doc.

linked to recommended or mandatory `HighSchoolMajors`, they belong to one or more `FieldsOfStudy`, and they are also related to `Keywords`. Importantly, EduKG comprises a total of 286 curricula. This means there are 286 semantically valid tail entities out of 5,452 entities when it comes to recommending university curricula to students.

KG20C [20] is a scholarly KG encompassing 5 different entity types: `Authors`, `Papers`, `Domains`, `Conferences`, and `Affiliations`. The goal is to recommend a conference where to publish a paper. For each triple of the form $(h, r, ?)$, we aim at retrieving the ground-truth tail t where h is an entity of type `Paper`, r accounts for the relation `publishedIn` and t is an entity of type `Conference`. An example would be retrieving the correct tail for the following test triple: (`LearningToEfficientlyRank`, `publishedIn`, ?). In this case, the ground-truth tail is `SIGIR`. Hence, for the S-TCNS strategy, only entities of type `Conference` are used to replace tails of positive triples during negative sampling. Compared to EduKG, the set of semantically valid tails for the relation `publishedIn` is much more limited: KG20C comprises 20 distinct conferences out of the 16,362 observed entities.

Table 1. Characteristics of EduKG and KG20C. $|\mathcal{E}|$, $|\mathcal{R}|$ and $|\mathcal{T}|$ stand for the number of entities, relations and triples, respectively. $|\mathcal{V}|$ denotes the number of semantically valid tails for the target relation.

Dataset	$|\mathcal{E}|$	$|\mathcal{R}|$	$|\mathcal{T}|$	$|\mathcal{V}|$
EduKG	5,452	27	36,301	286
KG20C	16,362	5	55,607	20

The number of triples in EduKG and KG20C is comparable. However, it should be noted that EduKG contains fewer entities but more relations than KG20C (see Table 1).

4.2 Experimental Setup

The experiments are conducted using a 5-fold cross-validation. In each cross-validation setting, the test fold only comprises triples whose relation r is the target relation. TransE, TransH, DistMult, and ComplEx are implemented using PyTorch. Importantly, when comparing different training strategies, models are instantiated with the same initialization seed. The choice of the number of epochs, negative samples, embedding dimensions, and learning rate are based on what was found to work best for these datasets. Regardless of the strategy used, we first perform 1,000 epochs of general training with early-stopping (B-RNS) to ensure that training is achieved in a reasonable amount of time and that the best evaluation metrics are recorded. 100 epochs of specialized training are subsequently performed with early-stopping by applying the two strategies S-RNS and S-TCNS. These two strategies are compared against the baseline

B-RNS. All models are trained using max-margin loss and Adam optimizer as in [12]. The hyperparameters are shared across all configurations to ensure a fair comparison of their respective performance: number of negative triples per positive one $C = 50$ (in accordance with [21]), embedding dimension $k = 20$, learning rate $\eta = 0.01$, margin $\gamma = 1.0$. For TransE and TransH, distance $d = L2$ was used. Due to the presence of 1-to-N, N-to-1 and N-to-N relations in EduKG and KG20C, negative sampling can still generate false-positive triples after the corruption phase. Consequently, such positive triples are filtered out before ranking. In this work, only filtered Hits@K are reported [18]. Finally, all entities observed in the KG (regardless of their type) are scored during evaluation so that Sem@K can be reported.

4.3 Results

The experimental results of the strategies outlined in Sect. 3.2 are presented in Tables 2 and 3.

Table 2. Evaluation results on EduKG. H@K and S@K stand for Hits@K and Sem@K respectively, for $K \in \{1, 5, 10\}$. Green cells indicate which strategy performs best for a given model and metric.

	B-RNS			S-RNS			S-TCNS		
	H@1	H@5	H@10	H@1	H@5	H@10	H@1	H@5	H@10
TransE	4.66	16.12	25.49	6.27	22.66	33.64	8.19	26.75	37.82
TransH	7.19	22.96	35.51	9.67	30.02	43.49	12.16	34.38	47.80
DisMult	5.40	18.35	29.11	5.75	19.91	31.55	6.06	21.70	34.03
ComplEx	5.14	17.91	26.62	7.28	21.27	34.12	7.10	25.14	38.08
	S@1	S@5	S@10	S@1	S@5	S@10	S@1	S@5	S@10
TransE	97.39	94.79	92.80	99.69	98.94	98.36	96.21	91.62	88.19
TransH	99.13	98.35	97.39	99.61	99.39	98.98	99.17	97.90	96.17
DisMult	95.77	95.23	95.02	97.17	96.75	96.45	97.04	96.44	96.01
ComplEx	82.31	79.47	77.80	98.69	97.94	97.61	98.48	96.99	95.39

Baseline Strategy Comparison Across Datasets. In this section, we focus on the performance of the baseline model (B-RNS) regarding Hits@K. Results achieved under B-RNS are presented in the top left parts of Tables 2 and 3.

Overall, Hits@K are higher on KG20C than on EduKG. At first glance, this may seem counter-intuitive as the number of distinct entities is three times higher in KG20C ($\sim$16K) than in EduKG ($\sim$5K) (see Table 1). In this case, the lower Hits@K values achieved on EduKG compared to KG20C may reasonably be explained by the number of relations in EduKG, which exceeds the number of relations in KG20C.

Table 3. Evaluation results on KG20C. H@K and S@K stand for Hits@K and Sem@K respectively, for $K \in \{1, 5, 10\}$. Green cells indicate which strategy performs best for a given model and metric.

	B-RNS			S-RNS			S-TCNS		
	H@1	H@5	H@10	H@1	H@5	H@10	H@1	H@5	H@10
TransE	16.21	33.48	41.58	27.33	62.97	76.08	16.37	33.84	41.92
TransH	44.56	78.99	87.29	46.50	84.14	94.51	47.55	80.89	87.41
DisMult	23.05	36.57	40.89	28.54	46.20	51.97	27.67	43.03	47.37
ComplEx	24.98	39.54	44.40	32.21	57.28	63.03	27.45	49.16	58.87
	S@1	S@5	S@10	S@1	S@5	S@10	S@1	S@5	S@10
TransE	43.90	27.15	20.59	93.68	83.59	72.62	36.91	21.95	16.26
TransH	97.42	86.46	69.87	99.82	99.26	93.54	96.15	75.36	54.81
DisMult	51.46	30.30	20.72	73.48	49.20	34.74	67.77	38.59	24.22
ComplEx	54.91	31.16	20.48	86.72	64.04	44.06	74.53	59.84	47.36

On both datasets, the best performing model under B-RNS is TransH and the worst performing one is TransE. The superiority of TransH over TransE is in accordance with [25]. In terms of Hits@K, DistMult is slightly better than ComplEx on EduKG and slightly worse than ComplEx on KG20C. Recall that ComplEx has been proposed to better model antisymmetric relations than DistMult [21]. EduKG does not contain any antisymmetric relation, which may explain the relatively similar Hits@K between DistMult and ComplEx. However, in KG20C, there is one antisymmetric relation (citedBy), hence the slightly better Hits@K provided by ComplEx compared to DistMult.

RQ1: Impact of the Training Strategies on Hits@K. In order to answer the first research question, we evaluate the impact of the specialization strategies on the recommendation performance, as measured by Hits@K. From a more general point of view, we note that S-RNS and S-TCNS consistently achieve better Hits@K compared to B-RNS, for all $K \in \{1, 5, 10\}$ and regardless of the dataset and KGE model used (see Tables 2 and 3).

Differences exist between S-RNS and S-TCNS. On EduKG, S-TCNS provides better Hits@K than S-RNS (on average +35.7% for S-TCNS and +22.8% for S-RNS on Hits@10, compared to the results achieved with the baseline B-RNS). The reason may come from the fact that S-TCNS puts even more emphasis than S-RNS on discriminating between valid candidates for replacement, as the negative sampling is type-constrained. On KG20C, the conclusion differs: S-RNS performs systematically better than S-TCNS (on average +62.0% for S-RNS and +12.3% for S-TCNS on Hits@10, compared to the results achieved with the baseline B-RNS). This could be attributed to the low number of semantically valid candidates in KG20C: as there are only 20 entities of type Conference, S-TCNS is limited to a narrow set of semantically valid entities when performing negative sampling. As such, training may quickly reach a stage where performance cannot be improved anymore because most negative triples stop providing any further guidance to the model on improving the embeddings. This echoes the zero loss problem [23] mentioned in Sect. 2.2. In our experiments, we indeed noted that under S-TCNS, early-stopping was usually triggered before S-RNS, especially for TransE and TransH. For TransE, early-stopping was triggered after 7 epochs in average under S-TCNS and after 70 epochs in average under S-RNS. For TransH, early-stopping was triggered after 4 epochs in average under S-TCNS and after 22 epochs in average under S-RNS.

Consequently, when the set of semantically valid candidates is relatively large, S-TCNS seems to be an appropriate strategy as it helps the KGE model focus on semantically valid triples only, which are harder to differentiate. When the set of semantically valid candidates is limited, S-RNS is expected to perform better by providing more diverse and numerous samples to learn from.

To sum up, although S-RNS and S-TCNS do not provide comparable performance gains, they both improve results in terms of Hits@K compared to the baseline strategy B-RNS. This directly answers **RQ1**: specializing training on a single relation actually enhances the performance of a pre-trained KGE model in a downstream recommendation task.

RQ2: Impact of Training Strategies on Sem@K. To answer the second research question, we evaluate Sem@K on all models with the three strategies B-RNS, S-RNS and S-TCNS. First, the impact of S-RNS is significant: this strategy achieves the best results across all configurations. Interestingly, the use of S-RNS may even allow weaker models to compete with the other ones in terms of Sem@K. For example, ComplEx achieves disappointing Sem@K on EduKG under the baseline B-RNS. But when using S-RNS, Sem@K values obtained with ComplEx are comparable to the other models.

Focusing on the impact of S-TCNS on Sem@K, we note that on both datasets, the combined use of this strategy with the two translational models TransE and TransH deteriorates the model ability to capture the semantic profile of the target relation, compared to the results achieved under B-RNS. On the contrary, using S-TCNS enhances Sem@K of DistMult and ComplEx, which are two semantic-matching models. As evidenced by the green cells in Tables 2 and 3 for Sem@K, $K \in \{1, 5, 10\}$, we clearly see that S-RNS consistently performs better under all configurations. This answers **RQ2**: even though integrating the entity type into the negative sampling (S-TCNS) can improve the ability of a model to capture the semantic profile of the target relation, it is always a less interesting option compared to randomly sampling tails among all entities (S-RNS).

It appears that evaluating the quality of a KGE model by only using rank-based metrics provides a limited view of its performance. In some cases S-TCNS gives better Hits@K than S-RNS. By analyzing the influence of type-constrained sampling on the understanding of the semantic profile of the target relation, we see that S-TCNS performs poorly – sometimes even worse than the baseline B-RNS. Therefore, Sem@K is complementary to traditional rank-based metrics and special attention should be paid to this metric whenever there is a need for predicting semantically valid entities.

5   Discussion

Although the literature often presents RNS as an ineffective sampling strategy for the generic link prediction task [8,23], we have shown it can be successfully used in a recommendation framework when combined with a specialization training procedure. In particular, S-TCNS sometimes improves Hits@K at the expense of the model understanding of the semantic profile of the target relation. In this case, there seems to be a trade-off. On the contrary, S-RNS appears to pursue both objectives at the same time: improving the ability of a KGE model to assign a high score to the ground-truth, and better retain the semantic profile of the target relation. By generating both nonsensical and hard negatives, S-RNS reports substantial improvement over the baseline B-RNS. This leads us to think that when it comes to recommendation, focusing on triples whose relation is the target relation (S-RNS) may actually lead to better results than both focusing and using an informed negative sampling method (S-TCNS). Although S-RNS is a simple strategy that does not require any filtering or additional information – contrary to S-TCNS that requires knowing the type of each entity – we clearly demonstrate it could significantly enhance the quality of a KGE model in a downstream recommendation task.

Jain *et al.* [9] investigate whether embeddings are actually able to capture KG semantics. Although our research questions differ, we show that in the recommendation task, specific training procedures may improve the ability of an embedding model to capture the semantic profile of the target relation. Our experimental results even highlight that models that initially perform poorly on Sem@K could achieve competitive results after proper specialized training.

Compared to the baseline, S-RNS improves both Hits@K and Sem@K on the target relation. However, our work did not study whether the two proposed specialization strategies decrease the overall quality of a KGE model, i.e. its performance in terms of Hits@K and Sem@K for other relations. In addition, we choose RNS as our baseline. This may have an impact on the performance gains of the two specialization strategies. One may wonder whether choosing a stronger baseline would impact the outcomes of this work. Finally, it should be noted that in both EduKG and KG20C, the target relation cardinality is N-to-1. It would be interesting studying whether our proposed specialization strategies perform similarly on target relations of other cardinality, especially N-to-N relations that are frequently observed in recommendation.

6   Conclusion and Future Directions

In this work, we approach the recommendation task as a link prediction task on a single target relation of a knowledge graph, i.e. the relation that represents the links between users and items to recommend. To this aim, we consider knowledge graph embedding models and evaluate their performance with three training strategies: one baseline strategy that trains a model on the whole graph, and two specialization strategies that refine this model by focusing on the target relation. These two specialization strategies differ in their negative sampling method. Beside evaluating our models with the usual Hits@K metric, we introduce Sem@K, a new semantic-oriented metric that reflects the semantic validity of the top-ranked candidate entities. In light of these two metrics, we clearly show that the specialization strategy which refines a pre-trained embedding model by randomly sampling negative tails for the target relation consistently enhances both Hits@K and Sem@K compared to the baseline. In future works, we will extend our analysis using a broader range of models (e.g. GNNs). We will also address multi-typed KGs, different application domains, and propose an adjusted version of Sem@K that would benefit from further theoretical guarantees.

References

1. Bordes, A., Usunier, N., García-Durán, A., Weston, J., Yakhnenko, O.: Translating embeddings for modeling multi-relational data. In: Conference on Neural Information Processing Systems (NeurIPS) 2013, pp. 2787–2795 (2013)
2. Cai, L., Wang, W.Y.: KBGAN: adversarial learning for knowledge graph embeddings. In: Proceedings of the 2018 Conference of the North American Chapter of the Association for Computational Linguistics, pp. 1470–1480 (2018)
3. Chowdhury, G., Srilakshmi, M., Chain, M., Sarkar, S.: Neural factorization for offer recommendation using knowledge graph embeddings. In: Proceedings of the SIGIR Workshop on eCommerce, vol. 2410 (2019)
4. Edwards, G., Nilsson, S., Rozemberczki, B., Papa, E.: Explainable biomedical recommendations via reinforcement learning reasoning on knowledge graphs. arXiv preprint arXiv:2111.10625 (2021)
5. Gao, C., Wang, X., He, X., Li, Y.: Graph neural networks for recommender system. In: Proceedings of the Fifteenth ACM International Conference on Web Search and Data Mining, pp. 1623–1625 (2022)
6. Grad-Gyenge, L., Kiss, A., Filzmoser, P.: Graph embedding based recommendation techniques on the knowledge graph. In: Proceedings of the 25th Conference on User Modeling, Adaptation and Personalization, UMAP, pp. 354–359 (2017)
7. Hajimoradlou, A., Kazemi, M.: Stay positive: knowledge graph embedding without negative sampling. arXiv preprint arXiv:2201.02661 (2022)
8. Islam, M.K., Aridhi, S., Smail-Tabbone, M.: Negative sampling and rule mining for explainable link prediction in knowledge graphs. Knowl. Based Syst. **250** (2022)
9. Jain, N., Kalo, J.-C., Balke, W.-T., Krestel, R.: Do embeddings actually capture knowledge graph semantics? In: Verborgh, R., et al. (eds.) ESWC 2021. LNCS, vol. 12731, pp. 143–159. Springer, Cham (2021). https://doi.org/10.1007/978-3-030-77385-4_9
10. Jain, N., Tran, T.-K., Gad-Elrab, M.H., Stepanova, D.: Improving knowledge graph embeddings with ontological reasoning. In: Hotho, A., et al. (eds.) ISWC 2021. LNCS, vol. 12922, pp. 410–426. Springer, Cham (2021). https://doi.org/10.1007/978-3-030-88361-4_24
11. Kadlec, R., Bajgar, O., Kleindienst, J.: Knowledge base completion: baselines strike back. In: Proceedings of the 2nd Workshop on Representation Learning for NLP, Rep4NLP@ACL, pp. 69–74 (2017)
12. Kotnis, B., Nastase, V.: Analysis of the impact of negative sampling on link prediction in knowledge graphs. arXiv preprint arXiv:1708.06816 (2017)
13. Krompaß, D., Baier, S., Tresp, V.: Type-constrained representation learning in knowledge graphs. In: Arenas, M., et al. (eds.) ISWC 2015. LNCS, vol. 9366, pp. 640–655. Springer, Cham (2015). https://doi.org/10.1007/978-3-319-25007-6_37

14. Lin, Y., Liu, Z., Sun, M., Liu, Y., Zhu, X.: Learning entity and relation embeddings for knowledge graph completion. In: Proceedings of the Twenty-Ninth AAAI Conference on Artificial Intelligence, pp. 2181–2187. AAAI Press (2015)
15. Liu, C., Li, L., Yao, X., Tang, L.: A survey of recommendation algorithms based on knowledge graph embedding. In: 2019 IEEE International Conference on Computer Science and Educational Informatization (CSEI), pp. 168–171 (2019)
16. Nickel, M., Tresp, V., Kriegel, H.: A three-way model for collective learning on multi-relational data. In: Proceedings of the 28th International Conference on Machine Learning, ICML, pp. 809–816 (2011)
17. Palumbo, E., Rizzo, G., Troncy, R., Baralis, E., Osella, M., Ferro, E.: Translational models for item recommendation. In: Gangemi, A., et al. (eds.) ESWC 2018. LNCS, vol. 11155, pp. 478–490. Springer, Cham (2018). https://doi.org/10.1007/978-3-319-98192-5_61
18. Rossi, A., Barbosa, D., Firmani, D., Matinata, A., Merialdo, P.: Knowledge graph embedding for link prediction: a comparative analysis. ACM Trans. Knowl. Discov. Data 15(2), 141–1449 (2021)
19. Sun, Z., Deng, Z., Nie, J., Tang, J.: Rotate: Knowledge graph embedding by relational rotation in complex space. In: 7th International Conference on Learning Representations, ICLR (2019)
20. Tran, H.N., Takasu, A.: Exploring scholarly data by semantic query on Knowledge Graph Embedding Space. In: Doucet, A., Isaac, A., Golub, K., Aalberg, T., Jatowt, A. (eds.) TPDL 2019. LNCS, vol. 11799, pp. 154–162. Springer, Cham (2019). https://doi.org/10.1007/978-3-030-30760-8_14
21. Trouillon, T., Welbl, J., Riedel, S., Gaussier, É., Bouchard, G.: Complex embeddings for simple link prediction. In: Proceedings of the 33rd International Conference on Machine Learning, ICML, vol. 48, pp. 2071–2080 (2016)
22. Wang, M., Qiu, L., Wang, X.: A survey on knowledge graph embeddings for link prediction. Symmetry 13(3), 485 (2021)
23. Wang, P., Li, S., Pan, R.: Incorporating GAN for negative sampling in knowledge representation learning. In: Proceedings of the Thirty-Second AAAI Conference on Artificial Intelligence, pp. 2005–2012 (2018)
24. Wang, Q., Mao, Z., Wang, B., Guo, L.: Knowledge graph embedding: a survey of approaches and applications. IEEE Trans. Knowl. Data Eng. 29(12), 2724–2743 (2017)
25. Wang, Z., Zhang, J., Feng, J., Chen, Z.: Knowledge graph embedding by translating on hyperplanes. In: Proceedings of the Twenty-Eighth AAAI Conference on Artificial Intelligence, pp. 1112–1119 (2014)
26. Weyns, M., Bonte, P., Steenwinckel, B., Turck, F.D., Ongenae, F.: Conditional constraints for knowledge graph embeddings. In: Proceedings of the Workshop on Deep Learning for Knowledge Graphs (DL4KG@ISWC), vol. 2635 (2020)
27. Wu, S., Sun, F., Zhang, W., Xie, X., Cui, B.: Graph neural networks in recommender systems: a survey. ACM Comput. Surv. (2020)
28. Yang, B., Yih, W., He, X., Gao, J., Deng, L.: Embedding entities and relations for learning and inference in knowledge bases. In: 3rd International Conference on Learning Representations, ICLR (2015)
29. Yang, Z., Ding, M., Zhou, C., Yang, H., Zhou, J., Tang, J.: Understanding negative sampling in graph representation learning. In: KDD '20: The 26th ACM SIGKDD Conf. on Knowledge Discovery and Data Mining, pp. 1666–1676. ACM (2020)
30. Zhang, Y., Yao, Q., Shao, Y., Chen, L.: Nscaching: Simple and efficient negative sampling for knowledge graph embedding. In: 35th IEEE International Conference on Data Engineering, ICDE, pp. 614–625 (2019)

Documenting the Creation, Manipulation and Evaluation of Links for Reuse and Reproducibility

Al Idrissou[1]([✉]), Veruska Zamborlini[2], and Tobias Kuhn[1]

[1] Vrije Universiteit, De Boelelaan 1105, 1081 HV Amsterdam, The Netherlands
{oid201,t.kuhn}@vu.nl
[2] Federal University of Espirito Santo, Vitoria, ES, Brazil
veruska.zamborlini@ufes.br

Abstract. Data integration is an essential task in the open world of the Semantic Web. Many approaches have been proposed that achieve such integration by linking related entities across data providers, but they lack the support for in-depth documentation of the involved processes such as the creation, manipulation and evaluation of links. As a consequence, detailed documentation that eases the understanding and reproducibility of underlying processes is needed for a reliable reuse of graphs of identity available in the Semantic Web. We present here an approach to document such links and their processes, building upon a representation we call VoID+. It enables link-publishers to provide data-users with information that better support them in accessing and using links. We show that our approach with the proposed VoID+ ontology allows us to address the relevant competency questions around the reuse of integrated Semantic Web data. We also demonstrate how our approach has been successfully implemented in the Lenticular Lens, a user interface tool that annotates links it discovers, manipulates or validates under user's guidance. Based on a real-life humanities case study, we can show that the ontology amply annotates links in its life-cycle for reliable decision making by data-users.

Keywords: semantic web ontology · semantic web vocabulary · ontology design · data integration · linkset

1 Introduction

Links connecting co-referent entities (a.k.a identity links[1]) constitute one of the pillars of Data Integration in the Semantic Web where entity matching techniques enable their discovery and *creation*. They are represented as RDF triples where the subject and object are described in one or more datasets. They can be grouped together in sets (or "graphs") and can be annotated with metadata such as matching scores to enable seamless navigation across datasets and hence increase the potential of addressing complex problems such as investigating innovation on the creative industries during the long Dutch Golden Age (Sect. 4.2).

---

[1] For readability, we use the terms "links" and "identity links" interchangeably.

© The Author(s), under exclusive license to Springer Nature Switzerland AG 2022
O. Corcho et al. (Eds.): EKAW 2022, LNAI 13514, pp. 81–96, 2022.
https://doi.org/10.1007/978-3-031-17105-5_6

As the quality of input-data, matching algorithms and discriminating criteria are not always perfect and the matching process is heuristic in nature, erroneous links may be introduced, forcing data-users to *evaluate* the links' quality prior to their usage. More importantly, for *reliable reuse*, one is expected to *assess* whether the context at hand fits the context in which selected links are discovered. A given link may originate from (i) the application of given methods under particular data filters and matching criteria or from (ii) *manipulation* (combination, intersection, ...) of other existing links created by different providers. At present, the context of the origins of links lacks a uniform and coherent method of representation.

Highlighting what was mentioned above, integrating data via links depends not only on the creation and/or manipulation of links but more so on their quality and the ability to understand their provenance. This motivates our argument that the potential for a set of links to qualitatively complement an interlinked dataset depends on the quality of its links which, in its turn, depends on the quality of the processes leading to their creation. Evaluating these processes starts with accessing the *metadata*, which is often lacking or not comprehensible enough. Indeed, most links in the Semantic Web are provided without much provenance, often embedded within datasets or as 'plain' sets of 'owl:sameAs' or 'skos:closeMatch'-like links [3]. Moreover, works meant for guiding or assessing link creation do not highlight the need for documenting it [1]. Ignoring this is an indulgence that may not always be worth it. Instead, we advocate that the reliance on metadata is of paramount importance to enable graphs of links to be reliably assessed, evaluated, reproduced, reused or queried in the search for interlinked datasets. So much so that [2] writes: "it is in the best interest of a dataset publisher to provide potential users of the data with information that supports them in accessing and using the dataset". Based on this, we investigate the following questions: *"how can we detail the documentation of the creation, manipulation and evaluation of links and hence enable the understanding and reproducibility of interlinked datasets so that they can be reliably (re)used?"*.

This paper aims to provide a means to a comprehensive semantic documentation of links and their processes. As such, the main contributions of this work are: (1) A conceptual overview of key elements to comprehensibly describe links and their processes; (2) VoID+, a concrete representation for the proposed elements that goes beyond the documentation of interlinked datasets; (3) in-depth insights into the effects of the proposed approach on a real-world use case.

Sections 2 and 3 present the state-of-the-art and the proposed approach. The latter is evaluated in Sect. 4 by proposing SPARQL queries addressing the competency questions, and presenting the results of their application to a real case study. The use case was developed and executed by humanities researchers in the Golden Agents project using the Lenticular Lens[2] tool, briefly introduced in Sect. 4, which implements VoID+, the proposed vocabulary. Section 5 presents some points of discussion while Sect. 6 concludes the article.

---

[2] https://lenticularlens.goldenagents.org/.

1.1 Competency Questions

At first glance, when a user comes across a set of discovered links, there are typically a number of nagging questions one wishes to answer for a reliable re-use, evaluation or reproduction. For example, one would like to know which sources and entities are covered, what algorithms and discriminating criteria are used, if the links are validated or if the resources are clustered. We propose here some competency questions for which a set of links should provide answers:

1. Given a set of links (Linkset) of interest. **(a)** What are the interlinked datasets involved? **(b)** If any, what sequences of restrictions (entity-types), property selections and value filter are applied to the entities of interest and how are the elements of a sequence of restriction combined? **(c)** What entity matching techniques (algorithms) are applied? If more than one is applied, how are they combined? **(d)** For a particular matching method, on which resource descriptions (property-values) are they applied; under which value-constraints and threshold?
2. Given a set of datasets and entity-types, what links are returned to the user if she is only interested in the ones that are: **(a)** Above a certain threshold? **(b)** Found by a specific method? **(c)** Validated as accepted? **(d)** Rejected above a certain threshold?
3. What set(s) of links is/are returned to a user interested in: **(a)** A certain dataset and/or entity type? **(b)** The use of a particular algorithm for link discovery? **(c)** A set of discriminating properties?
4. Given a set of links composed of multiple Linksets of interest: **(a)** What operators are used to generate the set of Linksets? **(b)** How are the operands combined? **(c)** How are each of the operands generated?

Few of the above questions can be answered using existing vocabularies but with limitations, for Example, 1 a/b and 2.b can be addressed using VoID [2] and/or Prov-O[3]. However, the level of details required to properly address each question is not achievable with current approaches.

1.2 Motivation

To support a comprehensible documentation of discovered links and their processes so that questions such as the ones above can be addressed, we highlight some motivating concepts area we intend to challenge. These are:

– **Data partitioning:** entities selected for matching purposes often do not constitute the whole data at hand, but a subset based on a specific type of entities but also on particular properties and their values. For example, only selecting from a university dataset AI students who had an internship.
– **Identity criteria:** the identity-link discovery process requires a set of rules or criteria for enabling the discovering links. These should be made explicit.

---

[3] https://www.w3.org/TR/prov-o/.

- **Multiple data sources:** a set of links may point to multiple data sources. Conceptually, nothing stops the source or target to be composed of multiple sources, contrary to what is observed in some vocabularies (see Sect. 2).
- **Multiple linking algorithms:** often, more than one type of metrics are used for object description comparison [6] only, no representation explicitly elaborates on their logical combination for reaching a combined link score.
- **Link-dataset manipulation:** as links can be produced under different settings/providers, the ability to manipulate those sets of links and their respectively computed matching scores using set-like operators may be informed.
- **Clustering:** the purpose of integration is to group co-referents[4], instances of heterogeneous sources, to address one or various pressing life problems.
- **Validation:** the reporting on the evaluation of not only the quality of links but also the quality of link processes enabling the mapping of digital representations to their unique counterpart in the sphere of real world objects.

To the best of our knowledge, at present, in the Semantic Web, no existing approach or vocabulary provides such detailed and broad coverage provenance on the integral processes surrounding the discovery of the links, their combination, and on whether and how they have been manipulated, clustered or evaluated.

2   Related Work

This section presents the related work on the VoID vocabulary and its limitations, on a number of approaches for representing links and their related processes, and on vocabularies used in the Semantic Web data integration tools.

2.1   VoID

The Vocabulary of Interlinked Datasets (VoID) [2], a standard advised by W3C, is a general purpose core vocabulary for providing metadata on datasets including graphs of links for the discovery and usage of interlinked datasets. However, it is not designed to dive into granular descriptions of specific datasets such as those only composed of links. Consequently, it is unable to inform data-users on *how links are generated and how the target datasets are semantically related.*

The observation is that it is (i) *a source of potential ambiguity in describing a partition* as it falls short of differentiating between a partition formed of entities that are for example [AIStudents and (had an internship or had an exchange program)] from one formed by entities that are [AIStudents or had an internship or had an exchange program] or in providing means to *restrict entities based on the value-range* of a property, for example, students born before the year 2000. Furthermore, for real world problems, it has a (ii) *too strict definition for a graph of links.* It forces instances of `void:Linkset` to be *directed* and to hold between *exactly two non-identical datasets* thereby explicitly disallowing a set of links to

---

[4] Co-referent is a term used in entity matching jargon to indicate a set of resources pointing to the same real-life object.

have more than one `void:subjectsTarget` or `void:objectsTarget`. This semantic does not allow, for example, for a linkset to hold for a dataset deduplication as it implies establishing links between duplicated resources stemmed from the *same dataset* with the intention of removing/merging redundant data.

2.2 VoID Extensions

The VoID ontology has seen a number of extensions over the years. **VoIDext** [5] is a vocabulary designed to enable the documentation of federated SPARQL queries in a way that highlights relatedness between datasets such that machines and humans can benefit. It proposes the concept *Virtual Links*, extending VoID with respect to querying links rather than detailing instance matching. **VoIDgen** [4] is designed for automating the description of large datasets using VoID by applying MapReduce paradigm to discover (sub)datasets. Besides reducing manual effort, incompleteness and inaccuracy, it proposes concepts such as Crisp versus Fuzzy linkset, enriching the semantics of datasets. **VoIDp** [12], is an ontology designed for the enhancement of interoperable datasets with virtual links.

Overall, the existence of extensions such as VoIDp, VoIDext and VoIDgen illustrate three interesting points crucial for the maturity of interlinked datasets. First, the extensions show the conformity with best practice, which advocates the *reuse of well-known vocabularies wherever possible*. Second, they exhibit the *acceptance of VoID* as a standardised core vocabulary for annotating interconnected datasets. Last but not least, they yet reveal limitations of VoID, hence the *need for new concepts* that best tackle the respective domains being modelled.

2.3 Vocabulary Used in Data Integration Tools

To a certain extent the metadata reporting on the links' provenance is addressed by some matching approaches / frameworks. SILK [15] provides a number XML-based files containing the matching specification. Only the resulting links are provided using RDF format, though it does not follow a known reification format for commenting on the identity links. It provides some means to use more than one matching method and combine the resulting scores using operators such as MAXIMUM, MINIMUM and AVERAGE. LIMES [10], another matching framework, have recently provided an RDF vocabulary called LIMES Configuration Ontology (LCO). However, as the name suggests, its main purpose is to express LIMES' linking-configuration in RDF. In the process, it uses VoID, but only specializing its main class, void:Dataset. It also provides a means to use more than one matching method and combines the resulting scores using operators such as AND, OR, MINUS and XOR. To the best of our knowledge, other approaches do not offer better means of documentation.

3   Approach: VoID+

The ontology here presented is called VoID+ as it is meant to extend and be compatible with the VoID vocabulary. Figure 1a provides a simplified overview of

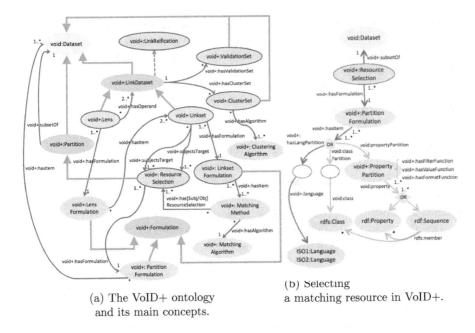

(a) The VoID+ ontology
and its main concepts.

(b) Selecting
a matching resource in VoID+.

Fig. 1. Linkset's method and context in VoID+

VoID+[5] highlighting (in bold-red-outline) its main elements: *Resource Selection,
Matching Method, Lens, Validation-set* and *Cluster-set*. In the sequel, these main
elements and their properties are further explained.

Resource Selection. This aspect concerns the selection of the resources under
scrutiny that can potentially end up being determined to be co-referent entities
during an entity matching process. Therefore, to perform a matching, one first
needs not only to select one or more data-sources, but also to restrict which
resources within each source will undergo the matching. The first way of doing
so is by applying a type restriction. Down this line, further restrictions can
be applied by forcing the value of a number of properties to lie within a certain
range. A Resource Selection is thereby the annotation of such a selection process.

In the excerpt depicted in Fig. 1b, for entity selection purposes we propose the
entity type `voidPlus:ResourceSelection`, which is a `voidPlus:Partition` based
on a `void:classPartition` and/or a `void:propertyPartition`. While the relation `void:classPartition` solely consists in specifying the type of entity under
scrutiny, the `void:propertyPartition` entails a little more. It consists in specifying a property or property path and a restriction that the selected property
should undergo for the selection of the right entities for the further down the
road entity matching process. Those restrictions can be combined using a formula description given by `voidPlus:hasFormulaDescription`.

---

[5] Details at https://lenticularlens.org/docs/03.Ontology/ and https://tinyurl.com/VoIDPlusGit.

Linkset Formulation For simple matching problems, finding co-referents can be done using a single matching algorithm. However, more than one is often needed for practical reasons. In this latter scenario, clearly reporting on how they work together for detecting co-referents is essential. As depicted in Fig. 2a, a Linkset Formulation entity is a resource for just doing the aforementioned.

Once resources of type Resource Selection are created, one can go ahead and use them for specifying the restricted collections to be used in a particular Matching Method. A resource of type `voidPlus:MatchingMethod` then specifies the Matching Algorithm and its arguments such as threshold, range and operator. In the end, all Matching Methods used in a matching process are documented using a resource of type `voidPlus:LinksetFormulation` which explicitly documents how they bind together in a logic expression given by the predicates `voidPlus:hasFormulaDescription` and `voidPlus:hasFormulaTree`.

Linkset. Linkset metadata (Fig. 2b) includes the WHO - WHAT - WHEN - HOW and related processes explaining the aboutness of links. While a Resource Selection entity specifies WHAT to match as subject and object targets, a Linkset Formulation specifies HOW entities are matched. Furthermore, some statistics on the matching results and other information can be reported such as the number of links found, the numbers of entities linked, WHO created the linkset and WHEN.

As discussed earlier in this section, according to the VoID documentation, the `void:Linkset` definition expects as data-sources exactly one source and one target, different from each other. This means it is more restrictive than the `voidPlus:Linkset` proposed here, since the latter also allows a linkset to connect resources *within* a single data-source or across more than two. As a consequence, we define a new concept rather than reusing `void:Linkset` in an incompatible manner. (`void:Linkset` is also not a subclass of `voidPlus:Linkset` as the latter requires the description of the processes underlying the creation of the links.)

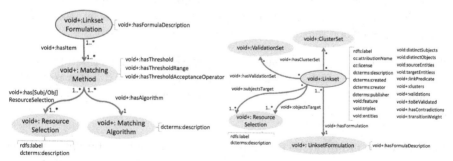

(a) Specifying the way in which methods are logically combined in VoID+.

(b) Specifying a linkset's context in VoID+.

Fig. 2. Linkset's method and context in VoID+

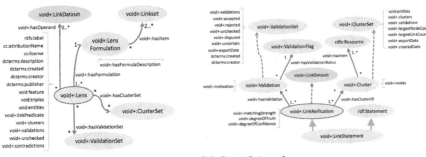

(a) Specifying the
Lens context in VoID+.

(b) Specifying the
Reification of Links in VoID+.

Fig. 3. Specification of a lens and a link Reification in VoID+

Lens. The creation of Lenses is another process that is important to document. In short, a lens is the result of a set-like operation — UNION, INTERSECTION or (SYMMETRIC) DIFFERENCE — over one or more Linksets and/or Lenses. For that, the entity `voidPlus:Lens` documents (Fig. 3a) its constituents as `voidPlus:hasOperands`. Like a `voidPlus:Linkset`, a `voidPlus:Lens` metadata includes `voidPlus:LensFormulation`, among others, and points to resources of types `voidPlus:Clusterset` and a `voidPlus:Validationset` if available.

LinkReification. The `voidPlus:LinkReification` resource, depicted in Fig. 3b, represents a reified link as a possible content of a `voidPlus:LinkDataset`. When reified, for example as an rdf:Statement using a standard reification or any other reification approach [7,9,11,13] for the matter, one can then annotate it with its own properties. For each link, VoID+ allows for one or more validations (`voidPlus:hasValidation`) by one or more users, belonging to one or more clusters (`voidPlus:hasClusterID`) created by different algorithms/processes. VoID+ does not enforce a particular reification type and link predicate, giving freedom of choice to the links creators. It requires, however, the users to be aware of the creator's choices when reusing or querying the linksets.

ClusterSet. One or more `voidPlus:ClusterSets`, using clustering algorithms, can be provided for a linkset or lens, which form a `voidPlus:LinkDataset`. This can support the validation process allowing for an overview of potentially equivalent entities.

ValidationSet: Qualitative Evaluation. When available, instantiations of one or more `void+:ValidationSet` is attached to a linkset or lens (`void+:LinkDataset`), comprising metadata with statistics and authority information on the validation process. Statistics on this matter can be included in the linkset metadata, particularly including eventual contradictions when more than one validation is provided, where for example some validations present a link as correct while other validation statements flag it differently.

4   Evaluation

The proposed vocabulary is evaluated by its ability to answer the proposed competency questions (Sect. 4.1), which are implemented as SPARQL queries and applied to a real case study. Due to space limitation, we only present two queries and their results while the remaining five queries and results are available online[6]. At present, only Question 2 cannot be fully answered as Item 2.b particularly challenges the proposed representation, as discussed in Sect. 5. In general, the complexity of the queries ranges from simple queries (4 patterns) to complex ones (recovering complex property paths used is a matching).

4.1   Queries For Competency Questions (QCQ)

Matching Results described using VoID+ help unveiling and understanding the sequence of processes leading to the creation, manipulation, clustering and validation of discovered links. Hereby, we present the queries addressing questions 1a & 1b and 2a & 2d.

QCQ 1.a & 1.b Given a linkset of interest, the query presented in Listing 1.1 addresses part of question 1 by (a) retrieving interlinked datasets and (b) their explicit partitions. This exhibits the selected class, properties and/or languages and how they are combined in a formula.

```
 1 SELECT DISTINCT ?PartitionLabel ?PartitionType
   ?Restriction ?PartitionInFormula ?filterFunction ?filterValue
 3 {
      <--LINKSET--> voidPlus:subjectsTarget|voidPlus:objectsTarget ?rscSelection.
 5
      # a) A dataset that is not itself a ResourceSelection
 7 ?rscSelection      voidPlus:subsetOf*            ?ds ;
                      rdfs:label                    ?PartitionLabel ;
 9                    voidPlus:hasFormulation       ?formulation .
   MINUS { ?ds a voidPlus:ResourceSelection . }
11
   ?formulation voidPlus:hasItem          ?partition   .
13 OPTIONAL { ?formulation voidPlus:hasFormulaTree ?PartitionInFormula . }
15 # b) Restrictions on the selected resources
   ?partition   a ?PartitionType;
17 { ?partition  void:class | voidPlus:language | void:property ?Restriction.
     FILTER (!isBlank(?Restriction)) }
19 UNION { SELECT ?partition
        (GROUP_CONCAT(DISTINCT ?propidx; SEPARATOR=" \n ") AS ?Restriction)
21    WHERE {
          ?partition void:property _:pseq .
23        _:pseq    a        rdfs:Sequence ;
                    ?seq      ?prop .
25      FILTER (?seq != rdf:type )
        BIND(CONCAT(strafter(str(?seq),"_"), " " ,str(?prop)) as ?propidx)
27    } GROUP BY ?partition  }
   OPTIONAL { ?partition   voidPlus:hasFilterFunction   ?filterFunction ;
29                         voidPlus:hasValueFunction     ?filterValue . }
   } ORDER by ?PartitionLabel  ?Restriction
```

Listing 1.1. Interlinked datasets & filters (1.a and b)

---

[6]   https://github.com/VoIDPlus-owl/EKAW2022/blob/main/CQs_for_VoIDPlus_EKAW.pdf.

QCQ 2.a & 2.d Given a set of data and entity-types of interest, Listing 1.2 retrieves links scored above 0.75 and rejected by a user in a validation process. Item (c) can be similarly addressed by selecting the accepted ones.

```
    SELECT DISTINCT ?linkDataset ?subDs ?sub ?objDs ?obj ?strength
 2  {
        VALUES ?givenDs { <--- SPECIFY THE DATASETS OF INTEREST ---> }
 4      VALUES ?givenType { <--- SPECIFY THE TYPES OF INTEREST ---> }

 6      ?linkDataset        voidPlus:hasOperand* /
            ( voidPlus:subjectsTarget | voidPlus:objectsTarget ) ?rscSelection .
 8      # Datasets Restrictions
        ?rscSelection       voidPlus:subsetOf+       ?givenDs ;
10                          voidPlus:hasFormulation ?formulation .
        # Type Restrictions
12      ?formulation        voidPlus:hasItem         ?typePartition  .
        ?typePartition      void:class               ?givenType .
14      # Standard Linkset Reification
        graph ?linkDataset { ?link    rdf:subject                 ?sub ;
16                                     rdf:object                  ?obj ;
                                       voidPlus:matchingStrength   ?strength . }
18      # Finding the dataset/entity-type selections for ?subj and ?obj
        ?linkDataset    voidPlus:hasOperand* / voidPlus:subjectsTarget /
20                      voidPlus:subsetOf*   /    rdfs:label               ?subDs ;
                        voidPlus:hasOperand* / voidPlus:objectsTarget /
22                      voidPlus:subsetOf*   /    rdfs:label               ?objDs .
        # 3.a links above a certain threshold (0.75)
24      FILTER (?strength > 0.75)
        # 3.d links that have been accepted
26      graph ?linkDataset { ?link voidPlus:hasValidation     ?v }
        graph ?validationSet { ?v  voidPlus:hasValidationStatus  resource:Rejected}
28  }
```

Listing 1.2. Links rejected yet above a preset threshold of 0.75 given a set of datasources and entity-types of interest. (2.a and d)

4.2   Use Case - Occasional Poetry

VoID+ is in use via the Lenticular Lens, a data-integration tool developed in the context of several projects[7], namely RISIS, CLARIAH and Golden Agents. The use-case discussed here is run over the latest implementation of the tool within the Golden Agents project, which uses the tool for *integrating data to engage in the investigation of complex problems that span the interaction between productions and consumption of the creative industries during the long Dutch Golden Age.*

The Lenticular Lens. It is a flexible tool[8] that aims at utilising generic off-the-shelves algorithms and a few tailored ones to allow the discovery of links across multiple datasets through users' guidance. Using the proposed VoID+ ontology, it allows detailed documentation of user's tailored matching processes and the full or partial export of these documentations in various reification flavors at the user's convenience. Not only does it support the documentation of links discovery but it also supports those of link manipulations, clustering and validations.

---

[7] http://risis.eu/, https://www.clariah.nl/, https://www.goldenagents.org/.

[8] The tool and the present use case implementation can found at:
https://lenticularlens.goldenagents.org/?job_id=90b598f72088ebd0e21446a12e353ffd.

Datasets. The Golden Agents project is working around the clock to make available to the public as many historical RDF datasets as possible by (i) converting humanities data into RDF and (ii) more than ever, enriching and linking the data of interest to the project. At present, it has in its Linked Open datasets portfolio a total of 27 and counting datasets of interest from 12 different content providers. The Lenticular Lens tool plays an important role since the project aims to encourage the addition of more and more RDF datasets to be interlinked, validated and reused by the public in the context or their own research.

As for the real life use case presented here, the following datasets are used:

- **SAA**[9]. The Amsterdam City Archives aka SAA, documents three social events that include *Trouw* (Marriage), *Doop* (Baptism) and *Begraaf* (Burial).
- **Occasional Poetry (Gelegenheidsgedichten).**[10] It contains metadata on poems that are among others written to celebrate the marriage of notables and describes poems with the name of the bride and groom, the marriage date, and bibliographic data such as information on the author and publisher.

Description. The use case aims at providing an enriched view on the *creation of poetry* back in the day. One possible outcome is to incite early start on literature by shedding light on the age in which poets started back then motivated by the occasion of, for example, the birth, marriage or death of notables. This can unveil relations between the type of festivities and honored persons and the poets and their ages. By connecting poetries with their corresponding event in SAA, it is likely to enrich biographical information on the honored persons and their social connections, since those datasets contain complementary information that can be used to help disambiguate mentions of people, mostly identified by their names.

Results. The finding of integration links is done using the Lenticular Lens. These links and their respective metadata are loaded into a Stardog 7.8 triple store so that they can be queried, for example, for generic competency questions translated into SPARQL queries in Sect. 4.1. As the metadata gets lengthy, it is not feasible to display it in the current article, but they can be observed by following the links given for the case-study. The use-case results in 4 linksets and 3 lenses, in a total of around 10K links between 14K entities. We present here the means to investigate under which conditions the links are created and combined, if they have been validated and more. Results are available online[11] as well as the corresponding links and metadata[12] which are exported from the tool. Hereby we describe an overview of the matches with statistics and present results of the queries addressing the competency questions.

---

[9] SAA: https://archief.amsterdam/

  This paper uses the RDF version of SAA data published by the Golden Agents project.

[10] https://www.kb.nl/bronnen-zoekwijzers/kb-collecties/oude-drukken-tot-1801/gelegenheidsgedichten-16de-18de-eeuw.

[11] https://lenticularlens.goldenagents.org/?job_id=90b598f72088ebd0e21446a12e353ffd.

[12] https://github.com/knaw-huc/golden-agents-occasional-poetry/.

Matching Overview. The Occasional Poetry dataset distinguishes *marriage* from *marriage anniversary*. For that and for independent analysis reasons, four linksets are created using the Lenticular Lens. Two that link notables to their notice of marriage (linksets 9 and 11) and two that link them to the baptism of their child (linksets 12 and 13). This then leads to the creation of three lenses: `lens:1` groups couples with the intention of marriage who got married within 6 months (`linkset:11`) with those celebrating at most 50 years of marriage anniversary (`linkset:9`); `lens:2` groups couples celebrating a marriage or a marriage anniversary to those baptising their child (`linkset:12` and `linkset:13` respectively); `lens:3` groups `lens:1` and `lens:2`.

Statistics. The integration modelled in Fig. 4 also highlights the total numbers of resources within each datasets respectively (Poetry:15,650, Marriage:1,197,673 and Baptism:4,889,160) and the number of links found per pair of datasets: 65,978 between Poetry and Marriage and 3,856 between Poetry and Baptism.

Fig. 4. Occasional poetry matching

CQ 1a & 1b inquire on (a) datasets involved and (b) the restrictions on them for a given link-dataset. Table 1, result of Listing 1.1, shows the metadata of `linkset:9`. It shows two dataset partitions: the City Archives' notice of marriage, in the 1st line, is partitioned simply based on the class roar:Person; the Occasional Poetry, in the next lines, is partitioned based on both the class schema:Person and on a property path that restricts those that are mentioned in a poetry about marriage anniversary (*jarig huwelijk*).

CQ 2a & 2d inquire on the links above a specified threshold that have been rejected given a set of datasets and entity types. Table 2, result of Listing 1.2, shows links with a score above 0.75 yet, listing matched resources (sub/obj), the resource selections from which they originate (subDs/objDs) and the strength resulting from the method applied. This example illustrates that passing the matching method's conditions does not necessarily mean that all resulting links are thereon valid. Often enough, such undesired contextual links need pruning and for that, other techniques such as [8, 14] can be applied for a more refined result.

5   Discussion

The development of VoID+ has triggered positive impacts in the humanities community within the Golden Agent project. It has facilitated the development and creation of the Lenticular Lens tool, which is also adopted as link discovery tool in other humanities projects. As the tool relies on VoID+ for its links'

Table 1. Results of QCQ 1a & 1b using Listing 1.1. It shows how datasets are partitioned for a given linkset, namely linkset:8f72088ebd0e21446a12e353ffd-9.

PartLabel	PartType	Restriction	PartitionInFormula	Filter	Value
SAA Notice of Marriage: Person	voidPlus: Class Partition	roar:Person			
OP: Person (marriage anniversary)	voidPlus: Class Partition	schema:Person	AND |- rsc:PropertyPartition-8a..24 |_ rsc:ClassPartition-a6..dd		
OP: Person (marriage anniversary)	voidPlus: Property Partition	1 inv(schema:about) 2 schema:Role 3 inv(schema:about) 4 schema:Book 5 schema:about 6 sem:Event 7 sem:eventType 8 sem:EventType 9 rdfs:label	AND |- rsc:PropertyPartition-8a..24 |_ rsc:ClassPartition-a6..dd	contains	%jarig huwelijk%

Table 2. Results of QCQ-2a & 2d partially displayed using Listing 1.2.
It shows links based on their strengths (> 0.75) and validation flags (rejected).

linkDataset	subDs	sub	objDs	obj	str.
lens:90..fd-3	OP: Person (marriage anniv.)	stcn: p067702015	SAA Notice of Marriage: Person	saa_deeds: 5e..5f?person=96..c98f..3b	0.84
linkset:90..fd-9	OP: Person (marriage anniv.)	stcn: p067763537	SAA Notice of Marriage: Person	saa_deeds: ab..c7?person=96..1efa..3b	0.88
lens:90..fd-3	OP: Person (marriage)	stcn: p067763537	SAA Baptism: Person	saa_deeds: ab..c7?person=96..1efa..3b	0.88
linkset:90..fd-9	OP: Person (marriage anniv.)	stcn: p069766037	SAA Notice of Marriage: Person	saa_deeds: 87..da?person=96..c29c..3b	0.76
. . .					

provenance, it ensures that all links discovered by researchers within the project's multiple real-life case-studies are properly and automatically documented. As a result, this allows researchers to now easily generate and consume links with more awareness of the context. However, there is still room for improvements, some of which are pointed out in this section.

VoID+ vs. OWL vs. SPARQL. This work has shown some of the limitations in the state-of-the-art's ontologies when it comes down to the documentation of links and their related processes. In particular, we emphasize here the advancement on complex partitioning offered by the proposed model as the state-of-the-art does not fit the bill. Even though some could argue to use OWL syntax in order to express restrictions such as conjunction, disjunction or property domain [5], we do not believe it is suitable for the description of partitions. That is because partitions are meant to restrict/select subsets of triples in a database, while OWL restrictions are meant to be applied over instances (individuals), eventually producing new triples in case those instances fit the restriction (e.g. class instantiation). Instead, the SPARQL-construct syntax allows exactly for selecting/describing a set of triples of interest based on a pattern (restrictions). In this scenario, VoID+ can be adapted to include the SPARQL-construct syntax.

Vocabulary and Granularity. Question 2.b poses an interesting knowledge representation challenge for link annotation: how much detail or what level of granularity is needed? As observed by [6], more than one metric is often used for resource description comparison in the link discovery process. In such scenario, in order to know which links result from the application of a particular method, one would need to annotate each link with the particular metric by which it is discovered. A rather drastic alternative as we see would be for linksets to be restricted to the use of a single method. However this solution imposes an efficiency issue since linksets simultaneously applying several methods allow for optimization in terms of time and space complexity.

6   Conclusion

Discovering, accessing and understanding the structure and schema of interlinked datasets are areas of metadata covered by VoID vocabulary among others. However, when it comes to (i) reproducing, (ii) understanding or (iii) assessing to reliably reuse datasets of triples interlinking other data, the state-of-the-art ontologies are presented with challenging issues as the concepts and semantics offered do not stretch to concepts that cover the aforementioned needs.

Based on important insights exhibited by competency questions and limitations of existing works ranging from generality, partition ambiguity, semantic restriction and limited concepts, we propose VoID+, an extension for VoID that enables creators, providers and users to document links with sufficient information to reliably support its reuse for context-based data integration.

For testing VoID+, we (1) illustrate generic queries that answer competency questions, (2) present the Lenticular Lens tool for link discovery in-use by projects like Golden Agents and Clariah, and which implements the proposed representation; and (3) discuss a use case of importance in the humanities through the Golden Agents project. With this real-life use case, we successfully show the importance of expanding the state-of-the-art vocabulary to main concepts related to the processes surrounding links in general. We also show how provenance can be extracted using SPARQL as illustrated in Sect. 4.1.

Overall, with the help of researchers within the humanities community and the Lenticular Lens, this paper shows (a) the importance of a full blown and more mature vocabulary for annotating links and their processes, (b) the need to help links creators to automate the annotation of discovered links. Put differently, the need for matching methods to self-document the links creation. Another issue shortly addressed in this work is (c) the usefulness of a flexible and generic tool for managing link-related issues such as the access to off-the-shelf matching algorithms; the generation, manipulation and validation of links.

Future Work. Enrich VoID+ with vocabularies such as PROV-O and SPARQL-construct. Investigate how to combine efforts with similar approaches such as SILK to improve the Lenticular Lens tool. Investigate the nature of computed scores and how to properly combine and manipulate them. For example, understanding during the computation of the final link-score what it means when SILK applies a weight per method to contextually distinguish the ones that are more relevant from the rest.

References

1. Albertoni, R., Pérez, A.G.: Assessing linkset quality for complementing third-party datasets. In: Proceedings of Joint EDBT/ICDT 2013 Workshops, pp. 52–59. ACM, New York (2013). https://doi.org/10.1145/2457317.2457327
2. Alexander, K., Cyganiak, R., Hausenblas, M., Zhao, J.: Describing linked datasets. In: Bizer, C., Heath, T., Berners-Lee, T., Idehen, K. (eds.) Proceedings of WWW 2009 Workshop on Linked Data on the Web, LDOW, vol. 538, pp. 10. CEUR-WS.org, Madrid (2009)
3. Beek, W., Raad, J., Wielemaker, J., van Harmelen, F.: sameAs.cc: the closure of 500M `owl:sameAs` statements. In: Gangemi, A., et al. (eds.) ESWC 2018. LNCS, vol. 10843, pp. 65–80. Springer, Cham (2018). https://doi.org/10.1007/978-3-319-93417-4_5
4. Böhm, C., Lorey, J., Naumann, F.: Creating void descriptions for web-scale data. J. Web Semant. **9**(3), 339–345 (2011)
5. Mendes de Farias, T., Stockinger, K., Dessimoz, C.: VoIDext: vocabulary and patterns for enhancing interoperable datasets with virtual links. In: Panetto, H., Debruyne, C., Hepp, M., Lewis, D., Ardagna, C.A., Meersman, R. (eds.) OTM 2019. LNCS, vol. 11877, pp. 607–625. Springer, Cham (2019). https://doi.org/10.1007/978-3-030-33246-4_38
6. Ferrara, A., Nikolov, A., Scharffe, F.: Data linking for the semantic web. Int. J. Semant. Web Inf. Syst. (IJSWIS) **7**(3), 46–76 (2011)
7. Hartig, O.: RDF* and SPARQL*: an alternative approach to annotate statements in RDF. In: Nikitina, N., Song, D., Fokoue, A., Haase, P. (eds.) Proceedings of ISWC, vol. 1963. CEUR-WS.org, Vienna (2017)
8. Idrissou, A., Zamborlini, V., Harmelen, F.V., Latronico, C.: Contextual entity disambiguation in domains with weak identity criteria, pp. 259–262. ACM (9 2019). https://doi.org/10.1145/3360901.3364440
9. Manola, F., Miller, E., McBride, B., et al.: RDF primer. In: W3C Recommendation , vol. 10, no. 1–107, p. 6 (2004)
10. Ngonga Ngomo, A.-C., et al.: LIMES: a framework for link discovery on the semantic web. KI - Künstliche Intelligenz, 413–423 (2021). https://doi.org/10.1007/s13218-021-00713-x
11. Nguyen, V., Bodenreider, O., Sheth, A.P.: Don't like RDF reification?: Making statements about statements using singleton property. In: Chung, C., Broder, A.Z., Shim, K., Suel, T. (eds.) 23rd International World Wide Web Conference, pp. 759–770. ACM, Seoul (2014). https://doi.org/10.1145/2566486.2567973

12. Omitola, T., Zuo, L., Gutteridge, C., Millard, I.C., Glaser, H., Gibbins, N., Shadbolt, N.: Tracing the provenance of linked data using void. In: Proceedings of International Conference on Web Intelligence, Mining and Semantics, WIMS 2011. Association for Computing Machinery, New York (2011). https://doi.org/10.1145/1988688.1988709

13. Orlandi, F., Graux, D., O'Sullivan, D.: Benchmarking RDF metadata representations: Reification, singleton property and RDF. In: 15th IEEE ICSC, CA, USA. pp. 233–240. IEEE (2021). https://doi.org/10.1109/ICSC50631.2021.00049

14. Raad, J., Beek, W., van Harmelen, F., Pernelle, N., Saïs, F.: Detecting erroneous identity links on the web using network metrics. In: Vrandečić, D., et al. (eds.) ISWC 2018. LNCS, vol. 11136, pp. 391–407. Springer, Cham (2018). https://doi.org/10.1007/978-3-030-00671-6_23

15. Volz, J., Bizer, C., Gaedke, M., Kobilarov, G.: Discovering and maintaining links on the web of data. In: Bernstein, A., et al. (eds.) ISWC 2009. LNCS, vol. 5823, pp. 650–665. Springer, Heidelberg (2009). https://doi.org/10.1007/978-3-642-04930-9_41

Should We Afford Affordances? Injecting ConceptNet Knowledge into BERT-Based Models to Improve Commonsense Reasoning Ability

Andrzej Gretkowski[1], Dawid Wiśniewski[1],
and Agnieszka Ławrynowicz[1,2]

[1] Faculty of Computing and Telecommunications, Poznan University of Technology,
Poznan, Poland
{dwisniewski,alawrynowicz}@put.poznan.pl
[2] Center for Artificial Intelligence and Machine Learning (CAMIL),
Poznan University of Technology, Poznan, Poland

Abstract. Recent years have shown that deep learning models pre-trained on large text corpora using the language model objective can help solve various tasks requiring natural language understanding. However, many commonsense concepts are underrepresented in online resources because they are too obvious for most humans. To solve this problem, we propose the use of affordances – common-sense knowledge that can be injected into models to increase their ability to understand our world. We show that injecting ConceptNet knowledge into BERT-based models leads to an increase in evaluation scores measured on the PIQA dataset.

Keywords: Commonsense reasoning · Natural Language Processing · Deep Learning · Knowledge Graph

1 Introduction

Equipping computers with the ability to understand the physical world is an important goal of artificial intelligence [2,8]. In recent years we moved closer to reaching it thanks to the rise of large pre-trained transformer-based models. These models may be taught using the language model objective, which requires them to learn to predict the next word in a given sequence or guess a masked word in a given text passage. Being trained over large textual corpora, these models learn world-related knowledge that helps them choose the right word.

However, a subset of knowledge called commonsense knowledge is not explic-itly stated in texts written by humans. Consider, for instance, a presupposi-tion [18] *a trolley is light enough to be capable of being pushed by a person* related to a statement *somebody pushed a trolley*. As some ideas are obvious to us and we expect that everyone is aware of them, we usually do not write about them. This problem is especially manifested regarding commonsense physical

© The Author(s) 2022
O. Corcho et al. (Eds.): EKAW 2022, LNAI 13514, pp. 97–104, 2022.
https://doi.org/10.1007/978-3-031-17105-5_7

knowledge on which we concentrate in this paper. This could be problematic, e.g., when using language models in embodied agents that need to interact in the physical world.

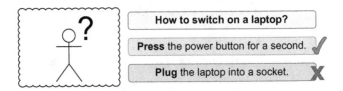

Fig. 1. An example of a question requiring affordances. We intuitively know that plugging the device into a socket is not enough to turn it on.

To address this issue, attempts to formalize commonsense knowledge are made. The promising idea is to use that formalized knowledge and inject it into pre-trained language models so that they can understand our world better. In this work, we utilize the notion of *affordances*, i.e., relationships between agents and the environment denoting actions that are applicable to objects, based on the properties of the objects (e.g., whether an object is edible, or climbable) [8] (Fig. 1). We extract knowledge about affordances from a knowledge graph to enrich the knowledge of popular pre-trained models. This paper's primary research question is *whether injecting commonsense knowledge concerning affordances into pre-trained language models improves physical commonsense reasoning.*

2   Related Work

Ilievski et al. [11] attempted to group commonsense knowledge into dimensions to verify which of them exactly impact models and concluded by stating that temporal, goal and desires are important dimensions for the models tested. On the other hand, a set of actions that a given object can make in a given environment are used in visual intelligence in the context of classification and labelling [28]. The authors focused on images, unlike our natural language-oriented work.

Commonsense reasoning in this paper is understood as an ability to make assumptions about ordinary situations humans encounter daily in their life. Among datasets that relate to this concept, there are ones that deal with multiple-choice questions [24,27]. In this paper, we use PIQA [2], which is a recent dataset focused on physical commonsense. The authors of the dataset prove that current pre-trained models struggle with answering questions collected in PIQA since they cover knowledge that is rarely explicitly described in the text (e.g., one has to choose whether a soup should be eaten using a fork or a spoon).

Some popular approaches to solve tasks requiring commonsense knowledge use GPT [7] or BERT-like models such as BERT [6], RoBERTa [14], ALBERT [12], or DeBERTa [10]. As they all follow the language model training objective, we expect they have some world-related knowledge. Results on PIQA using fine-tuned GPT model [3] achieved 82.8% accuracy. Fine-tuning such a model on another task seems to improve its performance consistently [19]. Recently, however, a DeBERTa-based model took the lead, achieving 83.5% accuracy on the leaderboard. There are also PIQA baselines based on BERT [2], but they score lower than DeBERTa and RoBERTa, which seem to be better when it comes to commonsense and overall performance on the aforementioned datasets, especially with highly optimized training hyperparameters [14]. It also appears that attention heads do capture the commonsense, which is encoded in graphs [5]. Moreover, UNICORN, a universal commonsense reasoning model trained on a new multitask benchmark using T5 (roughly 2 times bigger than BERT), where PIQA is a part, achieved 90.1% accuracy [15].

More specialized solutions include external resources that are used for fine-tuning or enriching the model output, such as graphs with labeled edges as interactions between actors [22] and relations between causes and effects [20]. Evaluations using such resources include inquiring a model for additional information [21] or combining data with graph knowledge in BERT models for classification [17]. There are also works that aim to re-define the distance between words using graphs [16] and generative data augmentation which seems to be a kind of adversarial training [26]. Recently, it was shown that adapter-based knowledge injection into BERT model [13] improves the quality of solutions requiring commonsense knowledge.

3   Affordances

The notion of *affordances* was introduced by Gibson [9] to describe relations between the environment and its agents (e.g., how humans influence the world). This relationship between the environment and an agent forms the potential for an action (e.g., humans can turn on a computer). Affordances help study perception as the awareness of the possibility to do certain actions related to the agent's world perception. As possibilities of actions – affordances – they are very natural for humans. This intuitively known knowledge may be underrepresented in internet-based textual corpora, while in some domains, such as robotics [1], one of the key reasoning tasks is inferring the affordances of objects (possible actions that can be accomplished with a given object at hand by a robotic agent).

For our use case, we can introduce several restrictions that may help to identify affordances: (i) *Affordance must explain some kind of relation* between two agents or concepts. This means it needs to touch on the aspect of how those two items coincide with each other or influence each other. (ii)*Affordance cannot be a physical connection.* Affordance is a metaphysical concept (a possibility of action) that connects two items. Thus, a cable connecting two computers is not an affordance. (iii) *Affordance cannot be a synonym.* While synonyms are connected by definition, affordance's goal is to explain how an agent connects to the counterpart in our world, not by just simply stating they mean the same. (iv) *Affordance cannot be a relationship based on negation.* There are many concepts out in the world that have some sort of relation. However, an affordance must in some way impact or be able to affect one of the agents.

4   Datasets

In this work, we use two datasets – PIQA and ConceptNet. PIQA, or "Physical Interaction - Question Answering", is a dataset of goals with two possible answers (further referenced to as solutions) provided. Only one of them is correct and choosing which requires some physical commonsense knowledge. For example, asking about how to eat a soup, our model should know that we want to use a spoon instead of a fork. PIQA is divided into train, validation, and test set.

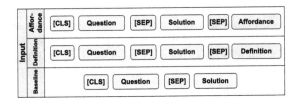

Fig. 2. Input differences between experiments in the architecture of the solution.

ConceptNet is a knowledge graph proposed to represent the general knowledge involved in understanding language, allowing applications to better understand the meanings behind the words [23]. It is based on data sources such as WordNet, OpenCyc, and Wikipedia. From all possible properties provided in the graph, we chose the ones that match the affordance requirements defined in Sect. 3. These are: *CapableOf*, *UsedFor*, *Causes*, *MotivatedByGoal*, *CausesDesire*, *CreatedBy*, *ReceivesAction*, *HasSubevent*, *HasFirstSubevent*, *HasLastSubevent*, *HasPrerequisite*, *MadeOf*, *LocatedNear*, and *AtLocation*.

5   Method

To inject the knowledge extracted from the ConceptNet graph, we need to identify appropriate subjects of the properties listed in Sect. 4 so that the objects related to a given subject via one of the selected properties may serve as an affordance. To achieve this goal, we extract keywords for each question and possible answers from PIQA using the tool YAKE [4]. The keywords found are then linked to ConceptNet. However, if no aforementioned subset of chosen properties is found in the context of a linked entity, we use a definition from the Wiktionary [25] as a fallback. The affordances selected are then passed to a model as part of an input representing a question and an answer pair. The affordance (or a definition from Wiktionary) is tokenized and placed after the last [SEP] marker following the input scheme: [CLS] QuestionTokens [SEP] SolutionTokens

[SEP] `AffordancesOrDefinitionsTokens`. Such an approach is in line with the original experiments with PIQA presented in [2], where similarly each question-solution pair is processed independently in the same manner and the embedding related to [CLS] token representing the whole context is processed by a single feedforward classification layer. We utilize the same approach simply adding affordances to the input so that the [CLS] token is aware of these (Fig. 2). With such a preprocessed input each of the base models is finetunned on the training set and then the results are obtained through the use of the validation set of PIQA. Preprocessing is done before the training begins and therefore it is the same on both sets of data.

6 Evaluation

We grouped affordances into 4 scenarios: (i) **standalone** aims to collect as many affordances as possible from all considered properties related to extracted keywords. These are then connected as sentences and added to the input as text. (ii) **just first**, extracts only the first affordance from a given keyword – the one that is the most important for the answer (meaning, we iterate by answer keywords first). (iii) **definition** adds affordances as well as Wiktionary definitions to the knowledge part of the input, merging both solutions. (iv) **complementary** aims to add definitions only when we lack any affordances, which is almost 87.4% of cases. This way, the number of separators in the input stays always the same but has either affordances or definitions given in the same place.

As PIQA provides a separate test set, we evaluate our classifiers on this subset using accuracy as a metric, which is a reasonable choice since the dataset is balanced (50% of examples should choose the first solution and the remaining ones the second one). We compared several popular BERT-based models, as they were proved to be good choices in the context of commonsense reasoning tasks. Some of them, like RoBERTa-large, are available on PIQA's leaderboard for comparison. However, we did not experiment with the top-ranked models like GPT-2 and DeBERTa since they consist of over 1.5B parameters, which makes them hard to fit into GPUs. Thus, we limit our research to popular baselines.

Table 1 provides a summary of accuracy for various models when **baseline** (no affordances), **definition**, and **affordance** scenario is concerned. As there are 4 possible affordances scenarios described above, here we report the scores obtained from the best scenario. Because each model was trained on Wikipedia being part of the training set, we can draw an interesting conclusion: adding definitions from Wiktionary (already seen in the training phase) impairs the overall performance of each model. Conversely, affordances seem to help the overall results on average, especially in cases of bad performance on the baseline, such as the ALBERT model – improving by almost 4%. Unlike the previous method, which seems to worsen the overall results, affordances might be a good way to inform the model about our physical world. In general, we see that injecting affordances is beneficial – in all tested models the accuracy increased.

An in-depth analysis of different types of affordances creation methods is summarized in Table 2. We can observe that the methods based on just the

Table 1. Model accuracy from three viewpoints: baseline – no additional knowledge, definition – knowledge from Wiktionary, affordance – affordances from ConceptNet.

Accuracy (%)	Baseline	Definition	Affordance
roberta-base	73.6	72.3	**74.4**
roberta-large	77.9	75.2	**78.9**
albert-base-v2	57.6	54.5	**61.6**
albert-xlarge-v2	57.9	52.3	**61.2**
distilbert-base-uncased	64.9	64.4	**66.9**

Table 2. Accuracy for various settings: Standalone – all possible affordances, Just first – only the first affordance found, Definition – all possible affordances and definitions, Complementary – only adding definitions when no affordances have been found.

Accuracy (%)	standalone	just first	definition	complementary
roberta-base	73.4	**74.4**	71.2	71.5
roberta-large	**78.9**	77.1	74.5	77.4
albert-base-v2	61.3	**61.6**	60.3	61.2
albert-xlarge-v2	**61.2**	53.4	53.8	54.4
distilbert-base-uncased	**66.9**	64.3	63.6	63.1

affordances seem to be better – for every model, one of the two methods that only use affordances obtains the highest accuracy. This observation solidifies the hypothesis that language models lack certain knowledge conveyed with affordances.

7   Conclusions

We investigated how language models respond to commonsense physical knowledge and how well they understand the subject. To this end, experiments were conducted to determine how the incorporation of commonsense knowledge into the input of the language model influences the results. This was contrasted with the normal encyclopedic definitions and results without any additional knowledge. To gain commonsense knowledge, this work introduces the concept of affordances to machine learning and answering questions using ConceptNet.

Different types of affordances were also looked at. The paper presents 4 different affordance injection methods with a description and implementation as well as a comparison between them. Surprisingly, they all lead to the same conclusion that the Wikipedia definition knowledge does not help the models to answer the questions – what is more, it usually even makes results worse. Of the methods tested in this paper, only those that rely solely on affordances are of value, namely the one that lists all possible affordances, and the one that lists only one, most important, affordance. These methods turned out to be the most effective in the generated experiments. We published the source code online[1].

Acknowledgement. This research was partially supported by TAILOR, a project funded by EU Horizon 2020 research and innovation programme under GA No 952215.

References

1. Beßler, et al.: A formal model of affordances for flexible robotic task execution. In: ECAI 2020, pp. 2425–2432 (2020). https://doi.org/10.3233/FAIA200374, https://ebooks.iospress.nl/doi/10.3233/FAIA200374
2. Bisk, Y., et al.: PIQA: reasoning about physical commonsense in natural language. In: Proceedings of AAAI, vol. 34, pp. 7432–7439 (2020)
3. Brown, T.B., et al.: Language models are few-shot learners. arXiv preprint arXiv:2005.14165 (2020)
4. Campos, R., et al.: YAKE! keyword extraction from single documents using multiple local features. Inf. Sci. **509**, 257–289 (2020)
5. Cui, L., et al.: Does BERT solve commonsense task via commonsense knowledge? arXiv preprint arXiv:2008.03945 (2020)
6. Devlin, J., et al.: BERT: pre-training of deep bidirectional transformers for language understanding. arXiv preprint arXiv:1810.04805 (2018)
7. Floridi, L., et al.: GPT-3: its nature, scope, limits, and consequences. Mind. Mach. **30**(4), 681–694 (2020)
8. Forbes, M., et al.: Do neural language representations learn physical commonsense? In: Proceedings of CogSci 2019, pp. 1753–1759. cognitivesciencesociety.org (2019)
9. Gibson, J.J.: The Theory of Affordances, Hilldale, USA, vol. 1, no. 2, pp. 67–82 (1977)
10. He, P., et al.: DeBERTa: decoding-enhanced BERT with disentangled attention. arXiv preprint arXiv:2006.03654 (2020)
11. Ilievski, F., et al.: Dimensions of commonsense knowledge. arXiv preprint arXiv:2101.04640 (2021)
12. Lan, Z., et al.: Albert: a lite BERT for self-supervised learning of language representations. arXiv preprint arXiv:1909.11942 (2019)
13. Lauscher, A., et al.: Common sense or world knowledge? Investigating adapter-based knowledge injection into pretrained transformers. CoRR abs/2005.11787 (2020). https://arxiv.org/abs/2005.11787
14. Liu, Y., et al.: RoBERTa: a robustly optimized BERT pretraining approach. arXiv preprint arXiv:1907.11692 (2019)
15. Lourie, N., et al.: UNICORN on RAINBOW: a universal commonsense reasoning model on a new multitask benchmark. In: Proceedings of AAAI, pp. 13480–13488. AAAI Press (2021)
16. Lv, S., et al.: Graph-based reasoning over heterogeneous external knowledge for commonsense question answering. In: Proceedings of AAAI, vol. 34, pp. 8449–8456 (2020)
17. Ostendorff, M., et al.: Enriching BERT with knowledge graph embeddings for document classification. arXiv preprint arXiv:1909.08402 (2019)

---

[1] https://github.com/AndrzejGretkowski/masters-piqa.

18. Potoniec, J., et al.: Incorporating presuppositions of competency questions into test-driven development of ontologies. In: Proceedings of SEKE 2021, pp. 437–440 (2021). https://doi.org/10.18293/SEKE2021-165
19. Rajani, N.F., et al.: Explain yourself! leveraging language models for commonsense reasoning. arXiv preprint arXiv:1906.02361 (2019)
20. Sap, M., et al.: ATOMIC: an atlas of machine commonsense for if-then reasoning. In: Proceedings of AAAI, vol. 33, pp. 3027–3035 (2019)
21. Shwartz, V., et al.: Unsupervised commonsense question answering with self-talk. arXiv preprint arXiv:2004.05483 (2020)
22. Speer, R., et al.: ConceptNet 5.5: an open multilingual graph of general knowledge. In: Proceedings of AAAI, vol. 31 (2017)
23. Speer, R., et al.: ConceptNet 5.5: an open multilingual graph of general knowledge. In: Proceedings of AAAI, AAAI 2017, pp. 4444–4451. AAAI Press (2017)
24. Talmor, A., et al.: CommonsenseQA: a question answering challenge targeting commonsense knowledge. arXiv preprint arXiv:1811.00937 (2018)
25. Wales, J.: The Wikimedia community: Wiktionary (2002). https://www.wiktionary.org/. Accessed 10 Oct 2021
26. Yang, Y., et al.: G-DAUG: generative data augmentation for commonsense reasoning. arXiv preprint arXiv:2004.11546 (2020)
27. Zellers, R., et al.: SWAG: a large-scale adversarial dataset for grounded commonsense inference. arXiv preprint arXiv:1808.05326 (2018)
28. Zhu, Y., Fathi, A., Fei-Fei, L.: Reasoning about object affordances in a knowledge base representation. In: Fleet, D., Pajdla, T., Schiele, B., Tuytelaars, T. (eds.) ECCV 2014. LNCS, vol. 8690, pp. 408–424. Springer, Cham (2014). https://doi.org/10.1007/978-3-319-10605-2_27

Towards a Knowledge Graph of Health Evolution

Alba Catalina Morales Tirado(✉)⬤, Enrico Daga⬤, and Enrico Motta⬤

KMi, The Open University, Milton Keynes, UK
{alba.morales-tirado,enrico.daga,enrico.motta}@open.ac.uk

Abstract. Electronic Health Records (EHR) contain detailed data of a person's health conditions and could provide emergency first responders with useful information. In previous works, we envisaged an intelligent system able to inspect health records and identify people in need of special assistance, by reasoning on the evolution of conditions over time. Unfortunately, there is a lack of resources regarding health condition evolution and recovery time. However, information available on the web could help in supporting domain experts for building a database of Health Condition Evolution Statements (HES). This paper addresses this knowledge gap and proposes a four-step methodology based on knowledge acquisition (KA) techniques that support the extraction of HES from public sources. The approach uses text classification algorithms and exploits SNOMED CT taxonomy to build a database of HES. More importantly, the proposed KA pipeline includes a human-in-the-loop model that captures knowledge from experts and ensures the construction of high-quality Knowledge Graphs (KG) to support the task at hand. We evaluate the approach with domain experts' help and discuss the user study results. Finally, we contribute the first curated Knowledge Graph of HES.

Keywords: Knowledge Graph · Health evolution · Condition evolution · Dataset · Knowledge Acquisition · Knowledge extraction · SNOMED CT

1 Introduction

Healthcare data use, particularly Electronic Health Records (EHR), has received increasing attention in recent years. EHR constitutes a valuable information asset for emergency support systems, they include extensive and fine-grained details of a person's medical issues. This information could provide a snapshot of people's health status. For instance, identifying vulnerable people or people otherwise requiring special assistance in the context of an emergency [7,8]. However, assessing ongoing health issues represents a challenge to first responders. EHR contain an overwhelming amount of information that emergency services cannot process effectively, for both its size and specificity; as a result, crucial data might be overlooked.

© The Author(s), under exclusive license to Springer Nature Switzerland AG 2022
O. Corcho et al. (Eds.): EKAW 2022, LNAI 13514, pp. 105–120, 2022.
https://doi.org/10.1007/978-3-031-17105-5_8

In previous work [8], we describe two knowledge components an intelligent system requires to reason on EHR automatically. The first component is the HECON Ontology (Health Condition Evolution Ontology) [9], a model for representing and reasoning on condition evolution over time. HECON defines the recovery process as a set of features called Health Evolution Statement (HES); whose components are *type*, *pace* and *duration*. Type refers to how the health condition evolves (e.g., improvement, decline, permanent, unaffected). Pace indicates the speed at which it changes (slow, moderate, fast), and Duration is an estimation of span (expressed as minimum and maximum range). The second element is knowledge about health condition evolution, specifically structured data that will support the annotation of conditions according to HECON. However, to the best of our knowledge, no existing structured data about condition evolution is available for reuse.

In this paper, we address this gap by designing a Knowledge Acquisition (KA) pipeline focused on extracting Health Evolution Statements (HES) from unstructured data sources. We expand the initial work presented in [8] by including a Human-in-the-loop (HITL) module. The HITL step uses the recommendations generated in previous steps to facilitate domain experts' tasks and annotate conditions with HES statements. By capturing domain experts' knowledge, we accelerate the annotation of health conditions and ensure the construction of a high-quality Knowledge Graph (KG). Contributions[1] of this paper are:

- A methodology that implements a Knowledge Acquisition pipeline for building a Knowledge Graph of Health Evolution.
- A tool that instantiates the HITL module.
- A user study with domain experts to (a) refine the HES KG and (b) evaluate the overall approach.
- The first database of health evolution information published as KG.

The rest of the paper is organised as follows. Section 2, describes the background and related work. Section 3 gives an overview of the proposed KA pipeline and Sect. 4 describes its implementation. In Sect. 5 we give details of the Knowledge Graph construction. In Sect. 6 we present the KA pipeline's evaluation and the user study's results. Finally, in Sect. 7, we summarise the conclusions.

2   Background and Related Work

In the Smart City environment, data is gathered by different means and from different sources to improve city services. In recent years, attention has focused on using Electronic Health Records to assist emergency services [1,7,8].

Data held in Electronic health records could reveal ongoing health issues and help us identify vulnerable people in need of assistance during an emergency. To illustrate this, consider a fire emergency in a large building. Among the people

---

[1] Repository: https://github.com/albamoralest/Health-Condition-Evolution-data base.

in the building, two require special assistance: one is a wheelchair user, and the other suffers from lung disease. Evidently, the wheelchair user needs support to evacuate, but such information is not typically known to emergency services. An intelligent system with access to the person's health records could identify a diagnosis of a "Fracture of the spine" (a permanent condition) and understand that such a condition does not improve over time. However, the second case is less obvious. A person suffering from a lung disease like "Obstructive bronchitis" might suffer acute symptoms (difficulty in breathing and walking) only when the diagnosis is recent. In fact, these symptoms may disappear with appropriate treatment over time. An intelligent system with enough information about health evolution could automatically evaluate this person's EHR, provide crucial information about their recent health issues and evaluate how severe they are at the time of the emergency.

Concerning EHR analysis, much of the literature is oriented to facilitate the visual representation of historical clinical data by reasoning on past health events [2,11]. However, existing literature [4] does not address the problem of identifying ongoing health issues by reasoning on health condition evolution from EHR.

In [8,9] we presented the work focusing on providing the knowledge components required by an intelligent system to represent and reason on health evolution. The HECON Ontology (Health Condition Evolution Ontology) [9] is the first model created with this objective. As described previously, HECON represents the recovery process as a set of features called Health Evolution Statement (HES). However, we identified a lack of structured data available for reuse.

As a result, in [8] we presented initial work on the automatic extraction of Health Evolution Statements (HES) from unstructured data sources. We collected text from public websites, such as NHS England[2] and MAYO Clinic[3] and linked this information to the SNOMED CT taxonomy. We used knowledge classification techniques such as Machine Learning to classify the collected text according to HES features. We used the identified HES and took advantage of SNOMED CT semantic features to propagate health evolution statements to a more significant number of SNOMED CT concepts. However, this approach had some limitations:

- The generation of HES was narrowed to the number of health conditions collected from public sources, typically the most common diseases. Instead, domain experts could provide knowledge of a larger number of health conditions.
- The methodology presented in [8] demonstrated that health condition evolution data supported the identification of vulnerable people effectively; nevertheless, it is imperative to evaluate the overall approach and the accuracy of the automatically generated HES.
- Although the recommended HES are generated using reliable and authoritative sources, the resulting data was not validated by domain experts.

---

[2] https://www.nhs.uk/conditions/.
[3] https://www.mayoclinic.org/diseases-conditions.

This paper describes the implementation of a four-phase Knowledge Acquisition (KA) pipeline. The pipeline uses health conditions text descriptions as input and returns a list of recommended HES. We complete the KA pipeline by including a Human-in-the-loop (HITL) step, enhancing the construction of the KG of Health Condition Evolution. We can infer from the limitations listed that integrating experts in the approach can benefit the overall process of automatically generating HES and that capturing their knowledge is a crucial step for ensuring the accuracy and quality of the information.

In order to validate the overall approach, we carried on a user study involving domain experts. The user study's objective is to evaluate the accuracy of the recommendations generated as part of the KA pipeline. Furthermore, we aim to assess the viability of annotating health conditions by incorporating a HITL step. Consequently, we sought to answer the following research questions:

- RQ1: Can health condition evolution statements be extracted automatically from descriptions in natural language?
- RQ2: Can we use ontological knowledge to derive new health condition evolution statements automatically?
- RQ3: How sustainable is the proposed methodology to populate a database of health evolution?

In what follows, we present the application of the proposed methodology for building a structured health condition evolution database and the results obtained from the user study that will answer the proposed research questions.

3   Methodology Overview

This section describes briefly the knowledge acquisition pipeline proposed to build a database of health evolution information (see Fig. 1). First, we collect the data that describes health condition evolution. Next, we apply text classification techniques and knowledge completion methods to extract recommended health condition evolution statements from natural language. Lastly, we use the recommendations and incorporate domain experts' knowledge to accelerate the annotation of health conditions and ensure a high-quality HES generation. In what follows, we summarise the steps of our proposed approach.

1. Corpus preparation. The first step of the pipeline is dedicated to identifying data sources that describe health evolution and preparing the corpus to be used in the next step. The sources should comply with characteristics such as: being an authoritative source, publicly available, extensive and including a description of health evolution. The aim is to collect text describing diseases, procedures, and conditions (e.g., asthma, appendicitis, bronchitis) and link them to the corresponding concept in SNOMED CT taxonomy. The final output is a corpus of health conditions organised by sentences.

2. Knowledge components extraction. The output from the previous step is a large corpus of sentences; however, only a few sentences provide information on the evolution of health conditions. Therefore, the next task is to identify and classify the sentences according to the HES components defined by HECON Ontology. We rely on Machine Learning techniques and develop a pipeline that includes the training and testing of a set of models [5] for each feature of the HES statement. Next, the best-performing models are used to predict a HES for each sentence in the corpus. Since a condition can have one or more recommended HES, the next task is to clean inconsistent and repeated HES. Lastly, we apply an algorithm that uses support and confidence as metrics to rank the most frequent combination of annotations. The output of this step is a collection of SNOMED CT concepts linked to one or more recommended HES.

3. Knowledge completion. The recommended annotations generated in the previous step have limited coverage of SNOMED CT concepts; therefore, we exploit the semantic structure of SNOMED CT taxonomy to find similar concepts that could share the same HES. Specifically, we use the SNOMED CT concepts' features to identify patterns and derive propagation rules. The rules expand the coverage of the HES to other concepts in SNOMED CT and make it possible to elicit a large dataset of SNOMED CT concept annotations.

4. Human-in-the-loop. Until now, the proposed methodology generated one or more recommended HES for each condition. Selecting the more accurate HES requires additional knowledge. Therefore, in this step, the objective is to capture domain experts' knowledge and build a more accurate database of health condition information. Domain experts contribute in two ways: (a) by assessing the recommended annotations and (b) by creating new ones. The final output is a curated database of health evolution statements.

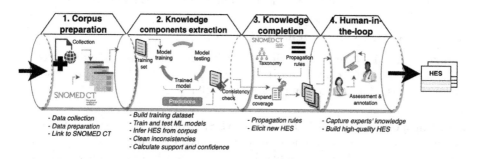

Fig. 1. Knowledge Acquisition pipeline

4 Knowledge Acquisition Pipeline

This section describes how we developed the proposed knowledge acquisition pipeline, as shown in Fig. 1.

4.1 Corpus Preparation

The first step is dedicated to identifying data sources that describe health condition evolution. The sources should come from (a) an authoritative organisation and (b) publicly available. Also, sources should be (c) extensive and (d) contain descriptions of condition evolution.

We identify two health websites: NHS England and MAYO Clinic. NHS England is the largest health website in the UK, providing straightforward access to content about symptoms, conditions, and treatments. The MAYO Clinic is a non-profit organisation; its website provides comprehensive and easy access to condition descriptions. NHS England website displays information of 972 health conditions and MAYO Clinic, 1170.

From both websites, we collect the HTML files that contain conditions descriptions. We clean the text by removing HTML tags, line breaks, special characters and empty spaces. After reviewing the text, we found out that condition evolution is usually described in one sentence. For instance, the evolution of "Broken ankle" is described as *"A broken ankle usually takes 6 to 8 weeks to heal, but it can take longer."*. Other conditions such as "Cataract surgery" has more than one description: *"It can take 2 to 6 weeks to fully recover from cataract surgery."* and *"These side effects usually improve within a few days, but it can take 4 to 6 weeks to recover fully."*. Therefore, we organise the corpus in sentences. The dataset contains 208,838 sentences in total, grouped by health conditions.

Typically, EHR uses SNOMED CT as a standard to describe clinical conditions [10]; therefore, we need to align the conditions' names from the web sources to SNOMED CT. These alignments facilitate the link of the HES to the EHR. We use Levenshtein distance to perform this alignment and find matching conditions' names. We run a manual review of the results in randomly selected conditions. The final corpus is a collection of sentences grouped by health conditions, where each health condition is linked to its corresponding SNOMED CT identifier.

4.2   Knowledge Components Extraction

Here, the focus is on extracting Health Evolution Statements candidate recommendations from the corpus (see Fig. 1, step 2). In what follows, we describe each task of the knowledge component extraction process in detail.

Building a Gold Standard Dataset of HES. In order to create this dataset, we examine the corpus. Health condition descriptions are extensive and have an average of 180 sentences. However, only a few of them describe health condition evolution; therefore, we use a distance supervision approach to identify these sentences. First, we select a sample of text snippets (expressions such as *last between, lifelong condition, no specific cure*, among others) that refer to condition recovery. Next, we use this sample and apply cosine similarity to automatically find a larger sample of sentences in the corpus. Then, we manually annotate each sentence with its corresponding HES components: type of condition (improve, decline, permanent), pace (fast, moderate, slow) and duration (maximum and minimum duration).

Finally, in preparation for the ML classification task, we add negative annotations to the training set. As mentioned earlier, the corpus contains sentences that do not express health evolution. We emulate this by adding sentences without this information and annotating them as "NONE". The output is a manually curated gold standard of 1987 sentences and their corresponding HES. Table 1 summarises the total number of sentences grouped by HES.

Table 1. Number of sentences per HES in the training dataset

Health Condition Evolution Statement (HES)			Total
Type	Pace	Duration	
NONE			1437
PERMANENT			141
IMPROVEMENT	MODERATELY	8 DAYS TO 2 MONTHS	106
IMPROVEMENT	FAST	5 MINUTES TO 1 DAY	74
DECLINE	SLOWLY	1 YEAR TO MORE YEARS	56
IMPROVEMENT	MODERATELY	2 MONTHS TO 6 MONTHS	53
IMPROVEMENT	FAST	1 DAYS TO 1 WEEKS	37
IMPROVEMENT	SLOWLY	1 YEAR TO MORE YEARS	37
IMPROVEMENT	SLOWLY	6 MONTHS TO 1 YEAR	30
DECLINE	FAST	1 DAY TO 1 WEEK	6
DECLINE	SLOWLY	6 MONTHS TO 1 YEAR	4
DECLINE	MODERATELY	8 DAYS TO 2 MONTHS	4
DECLINE	MODERATELY	2 MONTHS TO 6 MONTHS	2
TOTAL sentences			1987

Training and Testing Machine Learning (ML) Algorithms for the Classification Task. This task focuses on classifying sentences according to the dimensions used in HECON Ontology [9]. To perform this classification, we use the gold standard built previously as input for training and testing different ML algorithms [5].

We randomly divide the gold standard into training (70%) and test (30%) datasets, both with the same proportion of class labels as the gold standard. Table 2 shows the list of the ML algorithms we trained and the accuracy of each model grouped by HES features. Highlighted in bold are the models with the best performance. Also, hyper-parameters configuration for each algorithm can be found in the repository[4].

The training and test task is divided into two parts. For the first part, we perform a preliminary classification (C0). We train a boolean classifier to discriminate sentences that describe health condition evolution from those that do not contain such description. For instance, text such as *"Landing awkwardly from a jump"* is classified as NO (it does not describe health evolution) and *"There's currently no cure for chronic obstructive pulmonary disease (COPD)"* is classified as a description of health evolution. In this way, we aim to identify sentences containing condition evolution information and increase the number of true positives.

For the second part, we take into account the definition of the Health Evolution Statement. As described in [9], the HES comprises three features: type of health condition, pace and duration. Therefore, we use the sentences to train our classifiers across the three dimensions of the HES. For instance, the sentence *"There's currently no cure for chronic obstructive pulmonary disease (COPD)"*

---

[4] KG of Health Condition Evolution.

is used to train the different algorithms, first to obtain the type of condition, then the pace and the duration.

Table 2. ML training results: Accuracy per algorithm & HES features

ML Algorithms	C0	Direction	Pace	Duration
Logistic Regression	**0.8816**	**0.9727**	0.8148	0.8114
Decision Tree	0.8337	0.9272	**0.8934**	**0.8606**
Linear SVC	0.8789	0.9545	0.8271	0.8360
MLP Classifier	0.8337	0.9545	0.7530	0.6803
Naïve Bayes	0.4181	-	-	-
Multinomial NB	0.8136	0.8636	0.6790	0.5737
Random Forest Classifier	-	0.8818	0.7901	0.8442

Application of the Machine Learning Approach. Once we are satisfied with the results obtained in the previous task, we use the best-performing models for each feature of the HES (see Table 2) and make predictions on the entire corpus.

First, we run predictions using the best *"C0"* classification model. A total of 5,174 out of 208,838 sentences were classified as providing information about condition evolution. Table 3, column "Sentence", shows examples of sentences identified as positives. Second, we take this reduced dataset and run an independent classification process for each dimension of the HES. The first dimension is the type of health condition: improvement, decline or permanent. For instance, in Table 3, the sentence in the first row is classified as 'decline' and the one in the second row as 'permanent'. As described in HECON, only values "improvement" and "decline" have *progress* dimension. Thus, only 4,306 sentences, annotated as improvement or decline, were selected to complete the following two classification tasks (Pace and Duration). The example in Table 3 illustrates these cases. The sentence in the first row has a value for pace ('slowly') and duration ('1 year to more years'), unlike the sentence in the second row that is classified as 'permanent'. The output is a dataset of 5,174 sentences annotated according to the different features of HES model.

Consistency Check. In this task, our objectives are: (a) clean any inconsistencies that may arise from the classification and (b) produce metrics that allow the selection of the best HES among the recommended annotations. First, as some sentences were annotated with the same HES, we deleted repeated combinations of "condition + sentence + HES", leaving us with a total of 3,635 sentences. Next, we proceed to verify that the combination of features forms a coherent HES. For example, inconsistent combinations may have a pace annotation such as "fast" while duration indicates a long recovery 'from 6 months TO 1 year'. We rely on the pace and duration features to remove incoherent HES combinations.

The classification task generates one or more recommended HES; therefore, we provide metrics to select the best statement. We use an association rule learning method to identify how likely it is for a combination of HES features to represent a health condition. Firstly, we calculate how frequently the combination of health conditions and HES features appear in the dataset, and its support value (See Eq. 1). Then we calculate how often the combination of *health condition and HES is valid*, its confidence value as shown in Eq. 2. Table 3 shows the list of recommended HES for "Chronic obstructive lung disease" ranked by confidence. The ultimate output is a set of recommended Health Condition Evolution Statements (HES) linked to their corresponding SNOMED CT concept. A total of 1,324 SNOMED CT concepts have one or more recommended HES.

$$\{HES\} \Rightarrow Condition$$
$$\{Type, Pace, Duration\} \Rightarrow Condition$$
$$support = P(Type \cap Pace \cap Duration) = \frac{\text{number of predictions containing T, P and D}}{\text{total number of predictions}} \tag{1}$$

$$confidence(HES \Rightarrow condition) = \frac{supp(HES \cap condition)}{supp(HES)}$$
$$confidence = \frac{\text{number of inferences containing HES and condition}}{supp(\text{number of inferences containing HES})} \tag{2}$$

Table 3. HES best confidence value

SNOMED Concept	SNOMED Identifier	HES	Conf.	Sentence	Source
Chronic obstructive lung disease (disorder)	13645005	DECLINE SLOWLY FROM 1 YEAR TO MORE YEARS	0.0036	Although COPD is a progressive disease that gets worse over time, COPD is treatable.	MAYO
Chronic obstructive lung disease (disorder)	13645005	PERMANENT	0.0034	There's currently no cure for chronic obstructive pulmonary disease (COPD), but treatment can help slow the progression of the condition and control the symptoms.	NHS

4.3 Knowledge Completion

The data collected in Step Sect. 4.1 has limited coverage of SNOMED CT concepts. Therefore, in this task, we take advantage of SNOMED CT taxonomy and analyse the relationships and attributes of a given concept with the aim of finding similar concepts that could share the same HES. The objective is to identify patterns and create propagation rules, thus guiding an automatic HES expansion from SNOMED CT concepts with HES to other SNOMED CT concepts without HES, as illustrated in Fig. 1.

The logic model of SNOMED CT taxonomy includes components that represent two types of relationships [12]:

1. **Subtype relationship.** This is the most used relationship and is known as "*is a*" relationship or hierarchical relationship because they form the hierarchies in SNOMED CT. This means that the clinical detail of a concept increases with the depth of the hierarchies. For example, "Elbow fracture" → *is a* → "Fracture of upper limb".
2. **Attribute relationship.** This relationship contributes to the definition of the source concept by associating it with defining characteristics. The characteristics are called attributes and are specified by (a) the *relationship type* and

(b) the *value* provided by the destination of the relationship. For example, "Diabetes mellitus" attribute → is "Finding site (attribute)" and its value is → "Structure of endocrine system (body structure)".

Using these SNOMED CT taxonomy definitions, we follow a structured process to find patterns and derive generalised rules of propagation; in what follows, we enumerate the steps taken:

1. Manually select a source concept and analyse its features: the number of parents, attributes, and values.
2. Analyse if the features of the source concept (with HES) are shared by other concepts (without HES).
3. If identical or similar relationships (subtype or attributes) are found, then build a general query using SNOMED CT Expression Constraint Language (ECL) [13] and retrieve all concepts sharing the identified relationships.
4. Select a number of concepts from the results in the previous step and manually verify that the results share the same HES.

In what follows, we describe each of the rules created using the SNOMED CT features[5]. Table 4 presents an overview of the rules and examples.

Table 4. Propagation rules details.

Rules	General Rule	Example (ECL syntax)
Rule 1	All descendants of administrative related concept	<< 120646007 | Antibody screen (procedure) |
Rule 2	All immediate descendants of a source concept	<! 23406007|Fracture of upper limb|
Rule 3	All target concepts that share two or more attributes similar to source concept	(*):([1..1]363698007|Finding site (attribute)|= <<955009|Bronchial structure (body structure)|,116676008=4532008 AND ...
Rule 4	All target concepts with one attribute and direct children or the source concept	(102482005|Growing pains (finding)| OR <!102482005|Growing pains (finding)|):([1..1]363698007|Finding site (attribute)|= <<66019005|Limb structure (body structure)|)
Rule 5	All target concepts with same source parents OR source is parent AND similar attributes	((<<197480006|Anxiety disorder (disorder)|) OR <<21897009|Generalized anxiety disorder (disorder)|): ([1..1]363714003|Interprets (attribute)|=285854004|Emotion (observable entity)|)
Rule 6	All target concepts with two or more similar parents	(<!111273006|Acute respiratory disease (disorder)| AND <!32398004|Bronchitis (disorder)| AND <!128482007|Acute inflammatory disease (disorder)|)

- **Rule 1.** The hypothesis is that concepts describing administrative procedures do not affect people's health. For example, "Antibody screen" is an administrative procedure, thus, we annotated it as "UNAFFECTED". The same applies to its descendants.
- **Rule 2.** The hypothesis is that target concepts with a direct "is a" relationship inherit the source's HES. For example, "Elbow fracture" *is a* "Fracture of upper limb" and therefore inherits the source's HES.

---

[5] The term "source concept" is used to refer to a SNOMED CT concept that already has a HES annotation, and "target concept" to refer to a SNOMED CT concept that has no HES assigned.

– **Rule 3.** In this case, the target concept with two or more attributes similar to the source concept inherits the HES. For example, "Acute bronchitis (disorder)" shares its HES with concepts with similar attributes (e.g. Finding site, Associated morphology and Clinical course).

– **Rule 4.** Same as Rule 3, but restricted to one attribute descendants. For instance, "Growing pains" has one attribute "Finding site".

– **Rule 5.** In this case, a target concept with the same number of parents or similar attributes inherits the source concept HES.

– **Rule 6.** Here, a target concept with two or more similar parents as the source concept inherits the HES. This rule does not take into account concepts with one parent because the retrieved concepts are general.

The recommendations generated as a result of the classification and the knowledge completion task can be presented to experts to support the construction of the KG.

4.4 Human-in-the-Loop

In order to scale up the construction of the health evolution KG and build high-quality data, it is imperative to include domain experts in the loop; therefore, the last step of the pipeline focuses on capturing human knowledge (see Fig. 1).

This knowledge can be captured in three ways, by providing experts with (a) a list of recommended HES for each condition or (b) with a list of recommended target concepts that can share the same HES as the source concept; thus, they can assess the most accurate option swiftly. Also, experts can (c) build a new HES according to their best judgement.

We provide experts with a tool that reflects the options described above. The first interface displays the name of a condition and the list of candidate statements obtained in the knowledge components extraction step Sect. 4.2. Experts' task is to select the "Correct" HES according to their best judgement. The second interface uses the responses generated in the previous interface and the output from the knowledge completion step Sect. 4.3. The tool displays a condition (source), the HES that was selected as "Correct" in the first interface and the recommended conditions that could share the given source HES condition. When there is no recommendation available, experts can use a third interface and input a new HES using the different elements of the statement (type, pace, duration).

The final Knowledge Acquisition pipeline's output is a curated Health Evolution Statement (HES) database linked to its corresponding SNOMED CT concept. The KA pipeline is reproducible, and all the resources are available in the repository[6].

---

[6] Repository: https://github.com/albamoralest/Health-Condition-Evolution-database.

5   Knowledge Graph

In order to make the newly created database available in a structured and machine-readable format, we built a Knowledge Graph following the HECON Ontology model [9]. The Health Condition Evolution Ontology is a formal model representing the evolution of health events over time. Each HES in the curated database is linked to a SNOMED CT concept identifier; likewise, each SNOMED CT concept could be linked to one or more HES. The KG also stores data that represents the relationships between the data sources (MAYO Clinic and NHS England) and the process used to generate the annotation (knowledge component extraction, knowledge completion or HITL). This information supports the reasoning on the evolution of health conditions over time and the identification of ongoing health issues from EHR. The KG was built using SPARQL Anything [3], and it can be queried using SPARQL. Extended documentation, sample queries and the KG are available in repository[7].

6   Evaluation

In Sect. 4, we presented a complete knowledge acquisition pipeline to build a database of health evolution information. This pipeline included components that extract knowledge automatically from web sources and capture domain experts' knowledge. In what follows, we present the results of the user study carried out to evaluate the overall approach and answer the research questions stated in Sect. 2.

Evaluation Settings. To conduct the user study, we used the tool described in the Human-in-the-loop step 4.4. We invited medical students, interns, nurses, general practitioners, paramedics and first responders who are knowledgeable on how health events (medical procedures, health conditions, diseases) evolve. Seven people agreed to participate, each participant with a different level of expertise, as shown in Table 5. We divided the user study into two parts and adapted the HITL tool accordingly. For the first part, participants annotated the same randomly selected list of SNOMED CT concepts taken from the Knowledge components identification output. In the second part, participants annotated whether a target SNOMED CT concept shares the same HES as the source concept. For both parts, participants indicate whether the HES is correct or not using a five-category Likert scale: Incorrect, Partially correct, Neither correct nor incorrect, Partially correct and Correct.

---

[7] https://github.com/albamoralest/Health-Condition-Evolution-database.

Table 5. Participants by level of expertise

Total	Expertise	Current role	Specialisation
1	Research	EU project Officer	Palliative Care for Cancer Patients
1	Doctor	Trainee doctor	Psychiatry
2	Nurse	Nurse	Respiratory
		Nurse practitioner	Minor illnesses in a GP surgery
3	Student	Intern	Gynecology
		3rd year undergraduate student	Medicine
		3rd year PhD student	Clinical medicine research

6.1   Results

In what follows, we present the evaluation of the research questions formulated in Sect. 2.

RQ1. Can health condition evolution statements be extracted automatically from descriptions in natural language?

To answer RQ1, we (a) evaluate the feasibility of the task and (b) measure the accuracy of the recommendations. We present participants with the first interface of our HITL tool and ask them to annotate as many concepts as possible in 30 min. Participants had to indicate their level of familiarity with a given concept (familiar, partially familiar or unfamiliar) and whether the HES is correct or not using the five-category Likert scale. Also, they could input a new HES according to their best judgment.

Table 6, columns "Part 1", display the total number of annotations by participant. The results show that participants were able to use the recommendations and the HITL tool to annotate an average of 47 conditions in 30 min. Also, they generated, on average, seven new HES.

Turning now to the accuracy of the recommendations, the classification process generates one or more recommended HES per condition; therefore, to measure the number of relevant HES, we calculate Precision@k. It can be seen from the data in Table 7 that the system was able to provide useful recommendations in more than half of the cases (median precision@8 of 0.56). These results show that the extraction of HES is an achievable task.

RQ2. Can we use ontological knowledge to derive new health condition evolution statements automatically?

To answer RQ2, we use the recommendations generated by the knowledge completion step (see Fig. 1, step 3) and measure the number of annotations participants produce in 30 min. We provided participants with the second interface of our HITL tool and asked them to indicate whether a target SNOMED CT concept shares the same HES as the source concept. Similar to the evaluation in RQ1, participants should answer using a five-category Linkert scale. The source concept sample is constituted from the SNOMED CT concepts annotated as "Correct" and "Partially correct" in the first part of the study and the concepts for which the participants provided a new HES.

Table 6. RQ1 - Number of annotated SNOMED concepts per participant

	Part 1		Part 2	
	SNOMED concepts annotated	New HES generated	Source concepts	Target concepts annotated
P1	60	25	18	70
P2	47	0	32	143
P3	31	6	29	126
P4	67	0	62	284
P5	30	3	27	117
P6	53	6	37	162
P7	26	6	29	126
Avg	*45*	*7*	*33*	*147*

Table 7. RQ1 - HES Precision@k per participant

Precision @ k	1	2	3	4	5	6	7	8
P1	0.20	0.28	0.31	0.36	0.38	0.39	0.39	
P2	0.37	0.49	0.57	0.62	0.65	0.65	0.66	
P3	0.31	0.41	0.49	0.53	0.57	0.57		
P4	0.31	0.42	0.46	0.50	0.53	0.55	0.56	0.56
P5	0.26	0.30	0.37	0.42	0.47	0.49		
P6	0.28	0.34	0.39	0.42	0.42			
P7	0.23	0.30	0.38	0.38	0.38	0.38		
Median	*0.28*	*0.34*	*0.39*	*0.42*	*0.47*	*0.52*	*0.56*	*0.56*

As shown in Table 6, columns "Part 2", each participant was able to review an average of 33 source concepts and annotate 147 target concepts. In comparison with results in RQ1, where participants annotated an average of 45 HES, what stands out is that the exploitation of SNOMED CT taxonomy *produces three times (312%) more recommended HES*. Further analysis of the results shows that participants reviewed a total of 1,028 recommendations; these HES were annotated as "Correct" in half of the cases (501 conditions in total, see Table 8). These results demonstrate that the recommendations generated by the knowledge completion method are useful in half of the cases to swiftly populate the part of SNOMED that was not originally covered by the web sources.

Table 8. RQ2 - Total Correct annotations per participant

	Rule1	Rule2	Rule3	Rule4	Rule5	Rule6	Total
P1	13	0	4	4	6	0	27
P2	13	0	7	13	19	0	52
P3	9	0	19	11	28	3	70
P4	15	0	61	20	73	0	169
P5	19	0	16	14	32	4	85
P6	11	0	25	8	20	0	64
P7	1	0	10	4	18	1	34
Total	*81*	*0*	*142*	*74*	*196*	*8*	*501*
Proportion	*0.60*	*-*	*0.51*	*0.52*	*0.44*	*0.36*	*0.49*

Table 9. RQ1 - Total annotations by familiarity

	Familiar	P. familiar	Unfamiliar
P1	27	10	23
P2	9	19	19
P3	17	4	10
P4	33	22	12
P5	13	6	11
P6	14	17	22
P7	10	13	3
Total	*123*	*91*	*100*
Proportion	*0.39*	*0.29*	*0.32*

RQ3. How sustainable is the proposed methodology to populate a database of health evolution?

To answer RQ3, we analyse the results obtained in RQ1 and RQ2 and give an account of the effort (expressed in "person-month") required to populate the KG. We take as a reference the last edition of SNOMED CT, which included 353,567 concepts (published on January 31, 2020). On the one hand, in RQ1, one participant annotated an average of 50 correct concepts per hour (350 a day). If we only use recommendations generated by the knowledge component extraction step, it will take approximately four years and a half (55.20 person-months) to populate SNOMED CT. On the other hand, in RQ2, one participant annotated an average of 144 correct annotations per hour (1008 a day); it will take approximately a year and a half (19.16 person-months) to complete the task. We calculated only one person's effort, yet experts could perform the task simultaneously. For instance, with seven experts (emulating our user study) and the effort required in RQ1, the task will take approximately eight months (7.88 months). Likewise, the task is reduced to approximately three months (2.73 months) if considering the effort in RQ2 and seven participants. From these results, we can conclude that the approach is sustainable.

In addition, we evaluate the inter-rater reliability. We use Krippendorff's alpha coefficient [6] (applicable to missing data, various samples and different measures) and obtain an agreement of 0.4685. Although data in Table 9 indicate that participants were somehow familiar with 7 out of 10 concepts, the agreement result reflects the difficulty of finding participants with shared specialised expertise (as shown in Table 5).

7   Conclusions

In this paper, we presented a knowledge acquisition methodology to build a database of health evolution information. The pipeline implementation included automatic knowledge components such as text classification and completion. It also includes a Human-in-the-loop step to complete the methodology and obtain knowledge from domain experts.

The main goal of this paper was to fill the knowledge gap of resources regarding health condition evolution and recovery time. This study has found that extracting health evolution statements (HES) from natural language is possible. The results confirm that the recommendations facilitate the capture of knowledge from experts. Furthermore, exploiting SNOMED CT features accelerates the production of recommendations, hence the coverage of SNOMED CT.

A key strength of this research was the inclusion of the Human-in-the-loop module. The results of our user study show that including domain experts as part of the methodology accelerates the construction of the KG. More importantly, it ensures the capture of their valuable and accurate knowledge. With these results, we fill a gap in the literature and provide structured data on health evolution.

References

1. Abu-Elkheir, M., Hassanein, H.S., Oteafy, S.M.: Enhancing emergency response systems through leveraging crowdsensing and heterogeneous data. In: 2016 International Wireless Communications and Mobile Computing Conference (IWCMC). IEEE, September 2016. http://www.ieeexplore.ieee.org/document/7577055/
2. Alfattni, G., Peek, N., Nenadic, G.: Extraction of temporal relations from clinical free text: a systematic review of current approaches. J. Biomed. Inform. (2020). https://doi.org/10.1016/j.jbi.2020.103488

3. Daga, E., Asprino, L., Mulholland, P., Gangemi, A.: Facade-X: an opinionated approach to SPARQL anything. In: Further with Knowledge Graphs, vol. 53. IOS Press (2021). http://oro.open.ac.uk/78973/

4. Glicksberg, B.S., et al.: PatientExploreR: an extensible application for dynamic visualization of patient clinical history from electronic health records in the OMOP common data model. Bioinformatics **35**, 4515–4518 (2019). https://doi.org/10.1093/bioinformatics/btz409

5. Kotsiantis, S.B., Zaharakis, I., Pintelas, P.: Supervised machine learning: a review of classification techniques. In: Emerging Artificial Intelligence Applications in Computer Engineering, p. 20 (2007)

6. Krippendorff, K.: Computing Krippendorff's alpha-reliability (2011)

7. Morales Tirado, A.C., Daga, E., Motta, E.: Effective use of personal health records to support emergency services. In: Keet, C.M., Dumontier, M. (eds.) EKAW 2020. LNCS (LNAI), vol. 12387, pp. 54–70. Springer, Cham (2020). https://doi.org/10.1007/978-3-030-61244-3_4

8. Morales Tirado, A.C., Daga, E., Motta, E.: Reasoning on health condition evolution for enhanced detection of vulnerable people in emergency settings. In: Proceedings of the 11th KCAP Conference, pp. 9–16. ACM, December 2021

9. Morales Tirado, A.C., Daga, E., Motta, E.: HECON health: condition evolution ontology. In: 5th Workshop on Semantic Web Solutions for Large-Scale Biomedical Data Analytics, May 2022

10. NHS Digital services: National requirements for SNOMED CT. http://digital.nhs.uk/services/terminology-and-classifications/snomed-ct

11. Olex, A.L., McInnes, B.T.: Review of temporal reasoning in the clinical domain for timeline extraction: where we are and where we need to be. J. Biomed. Inform. **118**, 103784 (2021). https://doi.org/10.1016/j.jbi.2021.103784

12. SNOMED CT International: SNOMED CT Starter Guide (2017)

13. SNOMED CT International: Expression Constraint Language - Specification and Guide (2022). http://snomed.org/doc

Beyond Causality: Representing Event Relations in Knowledge Graphs

Youssra Rebboud$^{(\boxtimes)}$ ⓘ, Pasquale Lisena ⓘ, and Raphael Troncy ⓘ

EURECOM, Sophia Antipolis, Biot, France
{youssra.rebboud,pasquale.lisena,raphael.troncy}@eurecom.fr

Abstract. Dynamic environments can be modeled as a series of events and facts that interact with each other, these interactions being characterised by different relations including temporal and causal ones. These have largely been studied in knowledge management, information retrieval or natural language processing, leading to several strategies aiming at extracting these relationships in textual documents. However, more relation types exist between events, which are insufficiently covered by existing data models and datasets if one needs to train a model to recognise them. In this paper, we use semantic web technologies to design FARO, an ontology for representing event and fact relations. FARO allows representing up to 25 distinct relationships (including logical constraints), making it a possible bridge between (otherwise incompatible) datasets. We describe the modeling decision of this ontology resource. In addition, we have re-annotated two already existing datasets with some of the FARO properties.

Keywords: Semantic Web · Ontology · Event Relations

1 Introduction

In our experience of the world, we observe continuous occurrences of events. We may connect new events to one or more previous ones, giving birth to relationships of several types, such as cause-effect, relatedness, co-occurrence in time or space, etc. Even restricting our research to causality, we need to take into account several scenarios like *preemption* (causing an event which was going anyway to happen) and *disconnection* (making an event happening by removing the cause of not-happening) [23]. Events can influence each other (reciprocally or not), even without being recognised as cause-consequences. An event can be made of sub-events, each of them potentially relating to others. Being able to represent and exploit those relationships can be beneficial for different applications, involving the general public and domain experts.

The semantic web provides methods and tools to represent facts in Knowledge Graphs (KG) generally expressed in RDF. Some KGs are even specialised for representing event-centric information [7]. In Temporal Knowledge Graphs (TKG), each edge of the graph includes time information for identifying the

© The Author(s), under exclusive license to Springer Nature Switzerland AG 2022
O. Corcho et al. (Eds.): EKAW 2022, LNAI 13514, pp. 121–135, 2022.
https://doi.org/10.1007/978-3-031-17105-5_9

temporal validity of a triple [6, 20]. It is possible to use the event and time information for inferring edges [10, 27]. However, while TKGs are capable to represent an event occurrence, there are not suitable to represent inter-events relationships, making it hard to retrieve flow of events.

In this paper, we introduce the *Facts and Events Relationship Ontology (FARO)*, a data model for representing events relationships in Knowledge Graphs. In particular, we aim to design a structure which make possible to navigate through semantic links between events, exploring the flow of events backwards (searching for the causes or conditions of an event), forward (looking at consequences) or passing through other kind of connections. In other words, we want to make possible the creation of interconnected timelines of events, in which the connections between two consecutive points have explicit semantics. A such created graph would serve to improve the performance of downstream task (namely link prediction) and the explainability in decision making systems. We present several contribution:

- We compare a multitude of partially overlapping models, in order to understand which relationships should be represented because of interest of the community – Sect. 2;
- As an outcome of the literature review, we introduce the FARO model – Sect. 3;
- In other to foster future research, we realise a first Event Relation dataset that includes numerous event relations. This dataset has been obtained by re-annotating two existing datasets and by extending the TimeML format [22] with a new `RLINK` tag – Sect. 4.

We conclude in Sect. 5, summarising the contribution and the resources.

2   Related Work

In the literature, several works have studied event relationships, the most common type of relationships being **temporality**. Fan et al. [4] identified 13 temporal relations – to be used in the context of 3D simulation –, including simultaneity (*equal*) and 6 other asymmetric (directed) properties, with their respective inverse – e.g. *before/after*. Equivalent relations are included in [9], with the addition of *Vagueness*. **Mereology** in the context of events – i.e. the interaction between sub-events and super-events – is also often represented [6, 9, 13, 24, 29]. Finally, the literature mentions more kinds of relation that we can group under the name of **contingency**. Wolf distinguishes the causality relations in four different concepts [31]:

- CAUSE: event A that leads to an event B;
- ENABLE: condition C to make an event B possible;
- PREVENT: event A that avoids an event B;
- DESPITE: event A did not succeed in avoiding an event B.

Hong et al. [9] designed one of the most complete event-event relationship classification, including 5 types (Inheritance, Expansion, Contingency, Comparison, Temporality) and 21 sub-types, with possible overlaps between classes. To the best of our knowledge, this is the only work including **comparative relations** which cover three kinds of relation types, such as *Opposition*, when two events are improbable to be both true (*parole → sentenced*), *Negation*, when two events can be both true in different time slots, but not simultaneously (A is behind bars → A left). However, several relations between events are not accompanied by proper definitions, while still some relation types are missing.

Several ontologies have been published using semantic web technologies. While some of them do not include relations between events (e.g. LODE [25]), most of them include at least the concept of sub-events, such as in the *Event Pattern* [13], the *Event Ontology*,[1] and the *Simple Event Model (SEM)* [29].

Event Model F is an ontology created to support the response in emergency events [24]. It includes three kind of event relationships: mereological, causal and correlation. Its Justification class enables to support the relationship with provenance – e.g. opinion, scientific law, etc. However, this is modeled by including classes – e.g. EventCompositionSituation and EventCompositionDescription – with the only purposes of connecting events and defining their roles. As a consequence, there are no direct links between the composite super-event and its components sub-events (same for cause-effect). Furthermore, only 1-1 relations are foreseen, so additional instances must be created for aggregating causes/effects. All this led to a complex model, hard to understand and to adopt.

One of the most popular models among libraries and cultural institutions is *CIDOC CRM* [3]. It is an event-centric model, in which everything is represented though the interlinking of events of creation, production, movement, destruction, etc. Among its properties, there are some which intend or allow to interlink events, instantiating temporal relations (e.g. P176 starts before the start of), mereological relations (P9 consists of), causal relations (P17 was motivated by), and even include intentionality (P20 had specific purpose).

It is evident from the literature the necessity to represent, next to proper events, also some *state* or *condition*, lasting in time. This concept has been modeled as a sub-class of event [11] or as a completely separate class [5].

Several datasets for the detection of events and event relations are available, focusing mostly on temporal relations or on pure causality. Temporal relations have been largely investigated since 2009 in the TempEval shared task [28], which used the standard TimeML format and the TimeBank corpus [22]. The latter has been extended in CausalTimeBank [18] that follows the {CAUSE, ENABLE, PREVENT} model. In addition, events are marked as *factual* (happened), *counterfactual* (not happened) or *non-factual* (possibilities), while their relation can be *certain* or *uncertain*. On top of TimeML, the EventStoryLine dataset is proposed in [1], and includes the representation of causes and consequences in the context of PLOT_LINKs, for tagging events that are relevant in a plot.

---

[1] http://motools.sourceforge.net/event.

EventKG [6] is a knowledge graph of harmonised and interlinked events extracted from several resources, such as Wikidata and YAGO [8]. It includes over 1,3 million events, linked to their spatial and temporal coordinates. Only the connection between sub-events and super-events is represented in this dataset. For instance, it includes events such as *"Covid-19 lockdowns"* and *"Covid-19 pandemic in UK"*, with no direct relation between.[2] In the medical field, the datasets CSci [32] and EurekAlert [33] have been annotated according to four levels of causal relation: no relationship (c0), causal (c1), conditional causal (c2), and correlational (c3).

Table 1 summarises these models and datasets, showing which kind of relations are included in each of them. In addition to those, it is important to mention CausalNet, a common sense graph of actions, with weights between them indicating the likelihood that they are in a cause-effect relation [17]. Finally, it is worth to cite CausalBank – including 314 million sentence-level cause-effect pairs – from which it has been generated the Cause Effect Graph, in which links between events are weighted based on their co-occurrence in the text [15].

The table shows clearly that none of the existing resources is able to represent the entirety of the possible relations, calling for a more complete data model.

3   FARO: An Event Relation Ontology

Not all event relationships involve just events. For instance, one may want to describe that being tall is helping a player to score in a basketball game. The player's height is of course not an event, but rather a *condition* which supported the happening of an *event*. For this reason, FARO includes two different classes, *Condition* – transcendent, possibly can result in a RDF statement – and *Event* – immanent, following the categorisation in [23] – that are direct children of the more general class *Relata*, as in Fig. 1. The latter is not intended to be directly use for instantiate entities, but is rather an abstraction layer for the other two

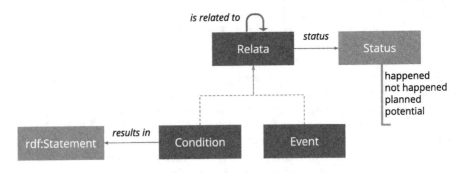

Fig. 1. Core elements of the FARO ontology

---

[2] We can logically imagine here that the spread the pandemic *caused* the lockdown, which is in its turn a measure for *preventing* the worsening of pandemic.

Table 1. Type of event relations which is possible to instantiate using certain schemas/ontologies or to find in certain datasets. In EventStoryLines (✔∗), it is possible to find plot actions which may be interpreted as cause of other events. In CSci and EurekAlert (✔+), it is also possible to express the conditional causation. In Causal TimeBank (✔-), both Enable and Prevent relations are separately considered in the process, but they are not distinguished in the dataset.

	Fan, 2008 [4]	Hong, 2016 [9]	CIDOC CRM [3]	EventModelF [24]	Event Pattern [13]	Simple Event Model [29]	Wolff, 2007 [31]	Event Ontology	TimeBank [28]	Causal TimeBank [18]	EventStoryLine [1]	EventKG [6]	CSci [32] / EurekAlert [33]	FARO
Temporal relations														
Before (after)	✔	✔	✔						✔	✔	✔			✔
Immediately Before (Immediately After)									✔	✔	✔			✔
Equal / Simultaneous	✔	✔	✔						✔	✔	✔			✔
Meets (is met by)	✔	✔	✔						✔	✔	✔			✔
Overlaps (is overl. by) / During	✔	✔	✔						✔	✔	✔			✔
Contains (is cont. by)	✔	✔	✔						✔	✔	✔			✔
Starts (is started by) / Begins	✔	✔	✔						✔	✔	✔			✔
Finishes (is finished by) / Ends	✔	✔	✔						✔	✔	✔			✔
Vague		✔												
Mereological relations														
Sub-event (super-Event)	✔	✔	✔	✔	✔			✔				✔		✔
Re-emergence	✔													✔
Coreference	✔													✔
Variation	✔													✔
Confirmation / Ev. type	✔					✔								
Contingent relations														
Cause	✔	✔	✔				✔			✔	✔∗		✔+	✔
Enable / Condition	✔						✔			✔-				✔
Prevent							✔			✔-				✔
Despite / Concession	✔						✔							✔
Correlation	✔				✔								✔	✔
Intention / Purpose				✔										✔
Not cause													✔	✔
Comparative relations														
Comparison	✔													✔
Conjunction / Similarity	✔													✔
Disjunction / Dissimilarity	✔													✔
Opposite	✔													✔
Negation / Alternative	✔													✔
Competition / Contrasting	✔													✔

main classes, allowing to define relations which connects indiscriminately any combination of them.

We found interesting to allow to define the *Status* of a *Relata* entity, to be chosen between four different options:

1. *happened* for sure at some moment in the past;
2. *not happened* for sure, we can exclude any happening of it in the future;
3. *potential*, meaning it is still uncertain if it will happen or not;
4. *planned*, sort of stronger potentiality, due to a will to this to happen.

This *Status* is intended to see an evolution in time, until it reaches either the *happened* or *not happened* status. We decided to leave possible to even define unforeseen statuses, apart to the four ones defined by the ontology.

Two *Relata* instances can be connected with a *is related to* property, which suggests general relatedness without further specification. The *is related to* property is further extended by 25 more specific properties, organised around four direct sub-classes of *is related to*, namely:

- comparatively related to
  - alternative to
  - compared to
    * dissimilar to
      · opposite to
    * similar to
  - contrasting version of
- contingently related to
  - causes
  - correlates with
  - does not cause
    * prevents
  - does not prevent (despite)
  - enables
  - intends to cause

- mereologically related to
  - coreference of
  - part of
  - re-emerges in
  - variation of
- temporally related to
  - before
    * immediately before
      · meets
  - contains
    * ends
    * starts
  - overlaps
  - simultaneous to

Differently from other works, we decided to structure these properties hierarchically, in order to enable reasoning. This hierarchy has been realised following the definition of the individual relations. For the same purpose, we included logic constraints – such as `owl:cardinality` and `owl:propertyDisjointWith` – and further define property characteristics – using `owl:SymmetricProperty` and `owl:Transitive Property`. Please note that FARO is only intended to be used for representing the relationships between events, leaving the event description to be represented using other vocabularies or ontologies.

Figure 2 shows two contingent relations, which represent using the FARO ontology the following text snippet: "A tight monetary program **caused** a temporary downturn but **prevented** a monetary meltdown".[3]

---

[3] The text sample has been taken from https://economynext.com/sri-lanka-will-repay-bonds-holders-should-appreciate-efforts-made-cabraal-83785/. Last visited: 10/06/2022.

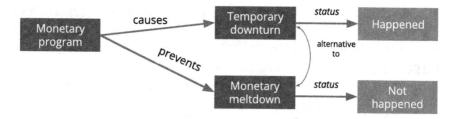

Fig. 2. A relata causing an event and preventing another one, represented using FARO.

Looking a second time at Table 1, it is possible to appreciate that FARO is covering most of the listed relations, proposing itself as central ontology for the harmonisation of different data models. We decided to not include in our ontology the *Vague* temporal relation: even if valuable from the point of view of information extraction, these kind of properties are not common in semantic web environments, where a more generic super-property can be used – in this case, *temporally related to*. Similarly, FARO is not including any *Confirmation/Event Type* property, because it can be expressed directly with an `rdf:type` statement. Alternatively, it is possible to use FARO in combination with other data models for event description – such as SEM [29], which allows typing events.

4   An Event Relation Dataset

In this section, we describe a dataset that includes some of the relations described in FARO, focusing on the contingent relations and in particular *Cause, Intend, Prevent, Enable, Not Cause*. The choice of targeting only a subset of the relations is due to time and resource constraints. However, we believe that a first version of a multi-relation event dataset is crucial to start designing new automatic methods for extracting them. Note that this is the first dataset incorporating *Intend*, and differentiating between *Cause, Prevent*, and *Enable*.

We developed this dataset by extending and re-annotating two existing datasets with new event relations types, namely intention, enabling, prevention, and explicit negation of causality. The choice of the datasets were based on their format (TimeML), which was convenient for extending it with other relation link.

– TimeBank [30], published by Brandeis University, providing 183 English news articles with over 27,000 event and temporal annotations about events, times and temporal links between events and times. The annotation respects the TimeML 1.2.1 specification.
– EventCausality [21], the dataset comes with causal and temporal annotations on 25 news articles collected from CNN7, giving at the end 1.3k events, 3.4k temporal links and 172 causal relations between events.

Both selected datasets are represented using the TimeML format [22], which we kept it as a base. This format enables to annotate events in the text and to declare possible connections between them using one among:

- TLINK, a temporal relation between events (or between an event and a time expression). Ex: "John **left (ei1)** 2 days *before (s1)* the **attack (ei2)**" →

  ```
  <TLINK eventInstanceID="ei1" signalID="s1"
  relatedToEvent="ei2" relType="BEFORE" magnitude="t1" />
  ```

- ALINK, a relationship between an "aspectual" event (events that add a notion about an action whether it begins, finishes, continues, etc.) – normally represented by phrasal verbs, e.g. *start to–* and its argument event: initiation, continuation, etc. Ex: John **started (ei5)** to **read (ei6)** →

  ```
  <ALINK eventInstanceID="ei5"
  relatedToEventInstance="ei6" relType="INITIATES" />
  ```

- SLINK, refers as a Subordination Link, which is used for contexts introducing relations between two events, or an event and a signal. Ex. "John **said (ei2)** that he **taught (ei3)** on Monday." →

  ```
  <SLINK    eventInstanceID="ei2"
  subordinatedEventInstance="ei3" relType="EVIDENTIAL" />
  ```

While TimeBank uses all 3 types of links, EventCausality instantiates explicit TLINK relation tags, with causal links are represented separately in another file – not following TimeML, so hard to re-use in other dataset. We kept the temporal links and we enriched it by new event relation tags.

4.1 A Generic Relation Link: RLINK

Following the experience described in [18] with the addition of the causal link CLINK, we extended TimeML with a new relation type RLINK, which we designed as a generalisation of the existing ones (TLINK, ALINK, CLINK), and enriched the previously described datasets accordingly. RLINK – or relation link – is a description of a generic relationship between two events, that can be further specified. A RLINK instance has 4 attributes as following:

- Link Identifier (lid) represents an ID for the relation, unique at the document level;
- Relation type (relType) refers as the type of relation between two events or the predicate of the triple, which can be one of the property of FARO, e.g. *Cause, Prevent*, etc.;
- Event instance Identifier (eventInstanceID) is the relata with the role of subject of the triple;
- Related event instance Identifier (relatedEventInstance) is the relata with the role of object of the triple.

Example. "Subcontractors will be offered a **settlement (ei264)** and a swift **transition (ei265)** to new management is expected to <u>avert</u> an **exodus(ei268)** of skilled workers from Waertsilae Marine's two big shipyards."

```
<RLINK eventInstanceID="ei264"
lid="142" relType="prevention" relatedEventInstance="ei268" />

<RLINK eventInstanceID="ei265"
lid="143" relType="prevention" relatedEventInstance="ei268" />
```

4.2   Candidate Generation

We re-annotated each of the mentioned datasets applying a semi-automatic procedure, based on expression matching as first step, followed by a manual check to validate the extracted annotations.

First, we collected a set of potential signal words for each of the 5 studied relations. We searched in the text these signals and extracted the sentences containing them, which we consider potential candidates. Each candidate sentence is dispatched according to the number of possible event pair combinations of relata that can construct the relation, among all the already annotated events for that specific sentence in the original datasets. In other words, we created a table in which each line contains a unique combination of two events, the signal word, the document id and the full sentence, as in Table 2.

Table 2. Table of the candidate pairs for a specific relation type (prevention), with manual annotation (1 = correct, 0 = wrong).

Event1	eid1	Event2	eid2	signal	Annotation	DocumentID	Sentence
settlement	e44	expected	e14	avert	0	$wsj_0187.tml$	Subcontractors will ...
...	...	...	...	...	0	$wsj_0187.tml$	Subcontractors will ...
settlement	e44	exodus	e46	avert	1	$wsj_0187.tml$	Subcontractors will ...
...	...	...	...	...	0	$wsj_0187.tml$	Subcontractors will ...
transition	e45	exodus	e46	avert	1	$wsj_0187.tml$	Subcontractors will ...

In the following, we detail the strategy applied for the signal collection and the extraction for each relation type, together with some examples.

Causality. We adopted the manually defined causal signals and causal verbs in [19], in which causal signals are nominal phrases that express causality (e.g.: because of, in order to, as a result of). However, causal verbs are a set of verbs representing the act of causing, such as: cause, bribe, push, etc. The first automatic selection results in 1790 candidate causal relation for TimeBank dataset, and 697 for EventCausality dataset. After dispatching, we ended up with 9658 and 1205 possible event pair causal relation for TimeBank and EventCausality datasets respectively.

Example. "Ocean Drilling amp Exploration Co. will <u>sell</u> its contract-drilling business, and <u>took</u> a \$50.9 million <u>loss</u> from discontinued operations in the third quarter **because** of the <u>planned</u> sale."

$$planned \xrightarrow{\text{causes}} sell \ , \ planned \xrightarrow{\text{causes}} took \ , \ planned \xrightarrow{\text{causes}} loss$$

Intention. To capture intention, we manually created a list of possible intention signals (e.g.: want, plan, aim). Additionally, we adopted another set of events as signals taken from the TimeBank dataset belonging to the class I-action. I-action (Intentional action), is an argument for those events that express an action of intention to do something.

Example. "Companies such as Microsoft or a combined worldcom MCI are **trying** to <u>monopolize</u> Internet access."
    The I-action is in bold face and, and the related event is underlined. The selection was manually performed after observing that some of these I-actions can alert the existence of this type of relation in a sentence.

Example. "Courtaulds PLC announced **plans** to <u>spin</u> off its textiles operations to existing shareholders in a restructuring to <u>boost</u> shareholder value."

$$spin \xrightarrow{\text{Intends to}} boost$$

As a result of automatic intention signals matching, we got 412 candidate expression for holding intention for TimeBank and 154 for EventCausality dataset. However, after extracting all possible event pair combinations, we ended up with 4028 and 230 intention candidate expression for TimeBank and Event-Causality datasets respectively.

Prevention. We integrate prevention signals as defined in [19], in which are initially included into the causal verbs list and claimed to express prevention, e.g.: block, bar, deter, etc. After the exploitation of these signals, we could extract 120 and 25 candidate expression, which lead to 988 and 53 event pair combination from TimeBank and EventCausality respectively.

Example. "In addition to the estimated 45,000 Marines to ultimately be part of Operation Desert Shield, Stealth fighter planes and the aircraft carrier John F. Kennedy are also <u>headed</u> to Saudi Arabia to **protect** it from Iraqi <u>expansionism</u>."

$$headed \xrightarrow{\text{Prevents}} expansionism$$

Enabling. For this event relation, we defined a list of verbs that alert the existence of enabling, such as authorize, warrant, entitle, etc.. We extended this list with enable signals as defined in [19], e.g.: help, permit, empower, etc.. to guarantee a high coverage. As a result, we obtained 41 and 17 candidate expression and 328 and 16 candidate event pairs combination for TimeBank and EventCausality datasets respectively.

Example. "In addition, Courtaulds said the <u>moves</u> are logical because they will **allow** the textile businesses to <u>focus</u> more closely on core activities."

$$moves \xrightarrow{\textbf{Enables}} focus$$

Not Causality. To extract the explicit not cause relation, we rely on the previously extracted causal relations, in which, we first naively pick those expression having both negation and causality at the same time, than manually validate the right ones. Consequently, we obtained 230 and 124 candidate expression and 1640 and 255 candidate event pairs from TimeBank and EventCausality datasets respectively.

Example. "He also rejected reports that his <u>departure</u> stemmed **from** <u>disappointment</u> the general manager's <u>post</u> had **not** also led to a board <u>directorship</u> at the London-based news organization."

$$disappointment \xrightarrow{\textbf{NOT cause}} departure \ , \ post \xrightarrow{\textbf{NOT cause}} directorship$$

4.3   Manual Annotation

The described process extracted a long list of candidate relations, most of them being incorrect and to be filtered out. The structure in Table 2 has been then used by two fluent English speakers annotators, which manually checked the candidate sentences. The process is summarised in the following steps:

1. Each annotator reads and annotates 300 lines for each type of relation.
2. On this preliminary annotation, we compute Cohen's kappa inter annotator agreement (IAA) [12] between the two annotations.
   - If the IIA does not show a substantial agreement (>0.6), the annotators meet, check the contrasting annotations and agree on a strategy. Then, 300 different lines are chosen and the process goes back to point 1.
   - Otherwise, we progress to next point.
3. The annotation is completed for the rest of the datasets, each annotator taking a unique portion.

During annotation, only relations with precised relata have been considered as correct, while others have been marked as not correct. The annotation process relied on an IAA = 0.7112, which is considered a substantial agreement.

In the following example, the signal word is marked in **bold**, events have been marked using *italic*, but only the underlined ones have been considered part of relationships of type *Prevent* by the annotators in Table 2.

Example. "Subcontractors will be offered a *settlement* and a swift *transition* to new management is *expected* to **avert** an *exodus* of skilled workers from Waertsilae Marine's two big shipyards, government officials *said.*"

4.4   Results and Discussion

Table 3 reports the number of candidate sentences, event pairs and final correct relations for each dataset and relation type, while in Table 4 we show the number of occurrences for each relation to have an insight about the balance of the final dataset, which can be useful in multi label classification tasks.

Table 3. Number of candidate sentences, event pairs and final correct relations for each dataset and relation type.

Dataset	Relation types	Extraction		Annotation
		n. of candidate sentences	n. of candidate event pairs	n. of correct relations
TimeBank	Cause	1790	9658	217
	Intend	412	4028	42
	Enable	41	328	11
	Prevent	120	988	17
	Not Cause	230	1640	3
EventCausality	Cause	697	1205	66
	Intend	154	230	2
	Enable	17	16	2
	Prevent	25	53	1
	Not Cause	124	255	0

Table 4. Total number of relations validated by annotators for each relation type. These relations are present in the released Event Relation dataset.

Relation type	Cause	Intend	Prevent	Enable	Not-Cause
Number of relations	283	44	13	18	3

Due to the applied strategy, we were able to only extract relations between events which have been explicitly tagged in the original datasets. This consequently affected the number of extracted links within each relation type, which is particularly low for *Not Cause*, *Prevent* and *Enable* – the subject of the latter not always being an event. The explicit negation of causality is not very expressed in the datasets that we covered, besides the native way of extracting them was not very efficient: indeed, collecting all sentences with causal signal and a negation has lead of lots of (false) candidate sentences.

5   Conclusion and Future Work

In this paper, we introduced FARO, an ontology for representing event relations. FARO includes a structured set of properties, which cover most of the relation types which can be found in the literature. In addition, we re-annotated two existing datasets in order to include some of the relation defined in FARO, releasing a new Event Relation dataset which can be used as ground truth for new multi-event extraction systems. FARO has been implemented in OWL and publicly documented.[4] The Event Relation dataset[5] is released in TimeML format. Both resources are published under an open source license.

We believe that empowering Knowledge Graphs with event relationship information will improve knowledge discovery and link prediction. At the same time, this kind of semantic representation can sensibly improve the explainability in decision making systems and the quality of text generation from graphs [26]. For this reason, we aim to realise a KG of events interconnected using semantically precise relations according to the FARO ontology, in which would be possible to follow relation chains and compute new ones. This KG should be populated by both extracting information from text and by interlinking with existing event-based KGs, such as EventKG [6] and YAGO [8].

In order to do it, an improved version of the Event Relation dataset should be realised. A first enhancement would come by offering a better coverage of different relation types. The used annotation methods can be improved, for example applying event detection techniques such as [2] in the candidate generation. We aim to use the annotated dataset within multi classification supervised tasks for event relation detection, in which we are considering the exploit and the adaption of previously implemented binary event relation extraction approaches – e.g. [14,16] – also enriching event representation with the involvement of common sense knowledge.

Acknowledgements. This work has been partially supported by the French National Research Agency (ANR) within the kFLOW project (Grant nANR-21-CE23-0028).

References

1. Caselli, T., Vossen, P.: The Event StoryLine Corpus: a new benchmark for causal and temporal relation extraction. In: Events and Stories in the News Workshop, Vancouver, Canada, pp. 77–86. Association for Computational Linguistics (2017). https://doi.org/10.18653/v1/W17-2711
2. Deng, S., et al.: OntoED: low-resource event detection with ontology embedding. In: 59th Annual Meeting of the Association for Computational Linguistics and the 11th International Joint Conference on Natural Language Processing, vol. 1, pp. 2828–2839. Association for Computational Linguistics, August 2021
3. Doerr, M.: The CIDOC conceptual reference module: an ontological approach to semantic interoperability of metadata. AI Mag. **24**(3), 75 (2003)

---

[4] https://purl.org/faro/.

[5] https://github.com/ANR-kFLOW/EventRelationDataset.

4. Fan, H., Meng, L.: Analysis of events in 3D building models. In: Liu, L., Li, X., Liu, K., Zhang, X., Chen, A. (eds.) Geoinformatics 2008 and Joint Conference on GIS and Built Environment: Geo-Simulation and Virtual GIS Environments, vol. 7143, pp. 1047–1058. International Society for Optics and Photonics, SPIE (2008)

5. Galton, A.: States, processes and events, and the ontology of causal relations. Front. Artif. Intell. Appl. **239**, 279–292 (2012)

6. Gottschalk, S., Demidova, E.: EventKG - the hub of event knowledge on the web - and biographical timeline generation. Semantic Web **10**(1039–1070), 6 (2019)

7. Guan, S., et al.: What is event knowledge graph: a survey. CoRR arXiv:2112.15280 (2021)

8. Hoffart, J., Suchanek, F.M., Berberich, K., Weikum, G.: YAGO2: a spatially and temporally enhanced knowledge base from Wikipedia. Artif. Intell. **194**, 28–61 (2013). https://doi.org/10.1016/j.artint.2012.06.001

9. Hong, Y., Zhang, T., O'Gorman, T., Horowit-Hendler, S., Ji, H., Palmer, M.: Building a cross-document event-event relation corpus. In: 10th Linguistic Annotation Workshop 2016 (LAW-X 2016), Berlin, Germany, pp. 1–6. Association for Computational Linguistics, August 2016. https://doi.org/10.18653/v1/W16-1701

10. Jin, W., Qu, M., Jin, X., Ren, X.: Recurrent event network: autoregressive structure inference over temporal knowledge graphs. In: Conference on Empirical Methods in Natural Language Processing (EMNLP) (2020)

11. Kaneiwa, K., Iwazume, M., Fukuda, K.: An upper ontology for event classifications and relations. In: Orgun, M.A., Thornton, J. (eds.) AI 2007. LNCS (LNAI), vol. 4830, pp. 394–403. Springer, Heidelberg (2007). https://doi.org/10.1007/978-3-540-76928-6_41

12. Kılıç, S.: Kappa test. J. Mood Disord. **5**(3), 142 (2015)

13. Krisnadhi, A., Hitzler, P.: A core pattern for events. In: 7th Workshop on Ontology and Semantic Web Patterns (WOP@ISWC), Kobe, Japan. IOS Press (2016)

14. Li, P., Mao, K.: Knowledge-oriented convolutional neural network for causal relation extraction from natural language texts. Expert Syst. Appl. **115**, 512–523 (2019). https://doi.org/10.1016/j.eswa.2018.08.009

15. Li, Z., Ding, X., Liu, T., Hu, J.E., Van Durme, B.: Guided generation of cause and effect. In: 29th International Joint Conference on Artificial Intelligence (IJCAI) (2020)

16. Liu, J., Chen, Y., Zhao, J.: Knowledge enhanced event causality identification with mention masking generalizations. In: 29th International Conference on International Joint Conferences on Artificial Intelligence, pp. 3608–3614 (2021)

17. Luo, Z., Sha, Y., Zhu, K., Hwang, S.W., Wang, Z.: Commonsense causal reasoning between short texts. In: Fifteenth International Conference on Principles of Knowledge Representation and Reasoning (KR), pp. 421–430. AAAI Press (2016)

18. Mirza, P., Sprugnoli, R., Tonelli, S., Speranza, M.: Annotating causality in the TempEval-3 corpus. In: EACL 2014 Workshop on Computational Approaches to Causality in Language (CAtoCL), Gothenburg, Sweden, pp. 10–19. Association for Computational Linguistics (2014). https://doi.org/10.3115/v1/W14-0702

19. Mirza, P., Tonelli, S.: CATENA: CAusal and TEmporal relation extraction from NAtural language texts. In: 26th International Conference on Computational Linguistics, pp. 64–75. ACL (2016)

20. Motik, B.: Representing and querying validity time in RDF and OWL: a logic-based approach. In: Patel-Schneider, P.F., et al. (eds.) ISWC 2010. LNCS, vol. 6496, pp. 550–565. Springer, Heidelberg (2010). https://doi.org/10.1007/978-3-642-17746-0_35

21. Ning, Q., Feng, Z., Wu, H., Roth, D.: Joint reasoning for temporal and causal relations. In: 56th Annual Meeting of the Association for Computational Linguistics, Melbourne, Australia, vol. 1, pp. 2278–2288. Association for Computational Linguistics, July 2018. https://doi.org/10.18653/v1/P18-1212

22. Pustejovsky, J., et al.: TimeML: robust specification of event and temporal expressions in text. In: Maybury, M.T. (ed.) New Directions in Question Answering, pp. 28–34. AAAI Press (2003)

23. Schaffer, J.: The Metaphysics of Causation (2016). https://plato.stanford.edu/archives/fall2016/entries/causation-metaphysics/

24. Scherp, A., Franz, T., Saathoff, C., Staab, S.: F-a model of events based on the foundational ontology DOLCE+DnS ultralight. In: 5th International Conference on Knowledge Capture (K-CAP), pp. 137–144. Association for Computing Machinery, New York (2009). https://doi.org/10.1145/1597735.1597760

25. Shaw, R., Troncy, R., Hardman, L.: LODE: linking open descriptions of events. In: Gómez-Pérez, A., Yu, Y., Ding, Y. (eds.) ASWC 2009. LNCS, vol. 5926, pp. 153–167. Springer, Heidelberg (2009). https://doi.org/10.1007/978-3-642-10871-6_11

26. Suchanek, F.: The need to move beyond triples. In: Text2Story - Third Workshop on Narrative Extraction From Texts (ECIR) (2020)

27. Trivedi, R., Dai, H., Wang, Y., Song, L.: Know-evolve: deep temporal reasoning for dynamic knowledge graphs. In: 34th International Conference on Machine Learning (ICML), vol. 70, pp. 3462–3471. JMLR.org (2017)

28. UzZaman, N., Llorens, H., Derczynski, L., Allen, J., Verhagen, M., Pustejovsky, J.: SemEval-2013 Task 1: TempEval-3: evaluating time expressions, events, and temporal relations. In: 7th International Workshop on Semantic Evaluation (SemEval), Atlanta, USA, pp. 1–9. Association for Computational Linguistics (2013)

29. van Hage, W., Ceolin, D.: The Simple Event Model, pp. 149–169. Springer, New York (2013). https://doi.org/10.1007/978-1-4614-6230-9_10

30. Verhagen, M., et al.: Automating temporal annotation with TARSQI. In: ACL Interactive Poster and Demonstration Sessions, pp. 81–84 (2005)

31. Wolff, P.: Representing causation. J. Exp. Psychol. General **136**, 82–111 (2007). https://doi.org/10.1037/0096-3445.136.1.82

32. Yu, B., Li, Y., Wang, J.: Detecting causal language use in science findings. In: 2019 Conference on Empirical Methods in Natural Language Processing and 9th International Joint Conference on Natural Language Processing (EMNLP-IJCNLP), pp. 4656–4666 (2019). https://www.aclweb.org/anthology/D19-1473.pdf

33. Yu, B., Wang, J., Guo, L., Li, Y.: Measuring correlation-to-causation exaggeration in press releases. In: 28th International Conference on Computational Linguistics (COLING), pp. 4860–4872 (2020)

Evaluating the Interpretability
of Threshold Operators

Guendalina Righetti[1]([⊠]) [iD], Daniele Porello[2]([⊠]) [iD],
and Roberto Confalonieri[1,3]([⊠]) [iD]

[1] Faculty of Computer Science, Free University of Bozen-Bolzano,
39100 Bolzano, Italy
guendalina.righetti@stud-inf.unibz.it
[2] Dipartimento di Antichità, Filosofia e Storia, Università di Genova,
16126 Genova, Italy
daniele.porello@unige.it
[3] Dipartimento di Matematica, Universitá degli Studi Padova,
Via Trieste, 63, 35121 Padova, Italy
roberto.confalonieri@unibz.it

Abstract. Weighted Threshold Operators are n-ary operators that compute a weighted sum of their arguments and verify whether it reaches a certain threshold. They have been extensively studied in the area of circuit complexity theory, as well as in the neural network community under the name of *perceptrons*. In Knowledge Representation, they have been introduced in the context of standard Description Logics (DL) languages by adding a new concept constructor, the Tooth operator ($\mathbb{W}$). Tooth expressions can provide a powerful yet natural tool to represent local explanations of black box classifiers in the context of Explainable AI. In this paper, we present the result of a user study in which we evaluated the interpretability of tooth expressions, and we compared them with Disjunctive Normal Forms (DNF). We evaluated interpretability through accuracy, response time, confidence, and perceived understandability by human users. We expected tooth expressions to be generally more interpretable than DNFs. In line with our hypothesis, the study revealed that tooth expressions are generally faster to use, and that they are perceived as more understandable by users who are less familiar with logic. Our study also showed that the type of task, the type of DNF, and the background of the respondents affect the interpretability of the formalism used to represent explanations.

Keywords: Threshold operators · Explainable AI · Interpretability · User study

1 Introduction

Predictive models based on machine and deep learning techniques have become ubiquitous in many decision making scenarios. Whilst these models are typically

This research is partially supported by Italian National Research Project PRIN2020 2020SSKZ7R and by unibz RTD2020 project HULA.

O. Corcho et al. (Eds.): EKAW 2022, LNAI 13514, pp. 136–151, 2022.
https://doi.org/10.1007/978-3-031-17105-5_10

very performative, they behave like black boxes, lacking transparency and leading to unfair and discriminative outcomes [23]. To this end, a lot of attention has been given to approaches that can explain black box models to increase trust by all users in why and how decisions are made [2,5,21].

Explainable AI (XAI) has been identified as a key factor for developing trustworthy AI systems [2,6]. The reasons for equipping AI systems with explanation capabilities are not only limited to enable diagnostics to prevent bias, unfairness, and discrimination [8], but also to user rights and acceptance (e.g., see Article 22 of the GDPR law [24]).

XAI focuses on developing approaches for explaining black box models by achieving good explainability without sacrificing system performance [18]. One typical approach is the extraction of local or global post-hoc explanations that approximate the behaviour of a black box model by means of an interpretable proxy. For instance, LIME is a local post-hoc explanation approach that explains model instances by means of linear expressions [26]. Other approaches advocate a tighter integration between symbolic and non-symbolic knowledge, e.g., by combining symbolic and statistical methods of reasoning [9,17].

Symbolic knowledge plays a key role for the creation of intelligible explanations. In [9], it has been shown that the integration of DL ontologies in the creation of explanations can enhance the perceived *interpretability*[1] of post-hoc explanations by human users. Furthermore, linking explanations to formal background knowledge brings multiple advantages. It does not only enrich explanations (or the elements therein) with semantic information—thus facilitating common-sense reasoning—, but it also creates a potential for supporting the customisation of the levels of specificity and generality of explanations to specific user profiles [19].

Motivated by the conventional wisdom that disjunctive normal form (DNF) is considered as a benchmark in terms of both expressivity and interpretability of logic-based knowledge representations [12], we assume to have local explanations of black box models modeled as a DNF formula. An example explanation from a loan agent could be: 'I grant a loan when the subject has no children and is married or when he has high income range' (i.e., $(\neg Parent \sqcap Married) \sqcup Rich$). Prior works raised the questions of whether DNF is always the most interpretable representation, and whether alternate representation forms enable better interpretability [4,12]. In particular, [4] evaluated several forms of DNFs in terms of their interpretability when presented to human users as logical explanations for different domains of application. In this work we aim at comparing the intepretability of DNFs and threshold operators.

Weighted Threshold Operators are n-ary operators which compute a weighted sum of their arguments and verify whether it reaches a certain threshold. These operators have been extensively studied in the area of circuit complexity theory (see e.g., [30]), and they are also known in the neural network community by perceptrons (see e.g., [3]). Threshold operators have been studied in the context of Knowledge Representation and integrated within DLs in [25], by adding a novel

---

[1] Interpretability describes the possibility to comprehend a black box model and to present the underlying basis for decision-making in a way that is understandable to humans [13].

concept constructor, the "Tooth" operator ($\mathbb{W}$). From now on, we shall use, more specifically "tooth operators" and "tooth expressions". Tooth operators allow for introducing weights into standard DL languages to assess the importance of the features in the definition of the concepts. For instance, as we shall see, the concept $\mathbb{W}^1((Parent, -1), (Rich, 2), (Married, 1))$ classifies those instances for which the sum of the satisfied weighted concepts reaches the threshold 1.

In the context of XAI, tooth expressions provide a powerful yet natural tool to represent local explanations of black box classifiers. In [14,16] a link between tooth-expressions and linear classifiers has been established, where it is shown that tooth-operators behave like *perceptrons*. More precisely, a (non-nested) tooth expression is a linear classification model, which enables to *learn* weights and thresholds from real data (in particular, from sets of assertions about individuals), exploiting standard linear classification algorithms. Thus, they could be used to represent post-hoc local explanations. Furthermore, adding tooth operators to any language including the booleans does not increase the expressivity and complexity of the language. Tooth expressions are indeed equivalent to standard DNFs,[2] i.e., canonical normal form of logical formulas consisting of a disjunction of conjunctions of literals [16]: they are 'syntactic sugar' for languages that include the booleans. They allow, however, for crisper formulas, being thus less error-prone and, putatively, more understandable by users.

In this paper, we present the results of a user study we conducted to measure the interpretability of tooth expressions versus their translation into standard DNFs. In the user study, respondents were asked to carry out different classification tasks using concepts represented both as a tooth-expressions and as DNFs. In line with previous works evaluating the interpretability of explanation formats (e.g., [1,4,9,10,20]), we used the metrics of accuracy, time of response, and confidence in the answers as a proxy for evaluating the interpretability of the two representations. We expected that tooth expressions could be perceived as more interpretable. In line with our hypothesis, our study revealed that the type of task, the background of the respondents, and the size of the DNF formula affect the interpretability of the formalism used.

2   Background

2.1   Tooth Operator - Preliminary Definitions

In this section, we delineate the formal framework necessary to introduce $\mathbb{W}$ (Tooth) expressions. Following the work done in [25], we extend standard DL languages with a class of m-ary operators denoted by the symbol $\mathbb{W}$ (spoken 'tooth'). Each operator works as follows: (i) it takes a list of concepts, (ii) it associates a weight (i.e., a number) to each of them, and (iii) it returns a complex concept that applies to those instances that satisfy a certain combination of concepts, i.e., those instances for which, by summing up the weights of the satisfied concepts, a certain threshold is met. More precisely, we assume a vector

---

[2] More precisely, non-nested tooth-expressions are not able to represent the XOR. Nested tooth can however overcome this difficulty.

of m weights $\vec{w} \in \mathbb{R}^m$ and a threshold value $t \in \mathbb{R}$. If $C_1, \ldots, C_m$ are concepts of $\mathcal{ALC}$, then $\mathbb{W}_{\vec{w}}^t(C_1, \ldots, C_m)$ is a concept of $\mathcal{ALC}_{\mathbb{W}}$. For C_i' concept of $\mathcal{ALC}$, the set of $\mathcal{ALC}_{\mathbb{W}}$ concepts is described by the grammar:

$$C ::= A \mid \neg C \mid C \sqcap C \mid C \sqcup C \mid \forall R.C \mid \exists R.C \mid \mathbb{W}_{\vec{w}}^t(C_1', \ldots, C_m')$$

To better visualise the weights an operator associates to the concepts, we often use the notation $\mathbb{W}^t((C_1, w_1), \ldots, (C_m, w_m))$ instead of $\mathbb{W}_{\vec{w}}^t(C_1, \ldots, C_m)$.

The semantics of $\mathcal{ALC}_{\mathbb{W}}$ just extends the usual semantics of $\mathcal{ALC}$ to account for the interpretation of the Tooth operator, as follows.

Let $I = (\Delta^I, \cdot^I)$ be an interpretation of $\mathcal{ALC}$. The interpretation of a $\mathbb{W}$-concept $C = \mathbb{W}^t((C_1, w_1), \ldots, (C_m, w_m))$ is:

$$C^I = \{d \in \Delta^I \mid v_C^I(d) \geq t\} \tag{1}$$

where $v_C^I(d)$ is the *value* of $d \in \Delta^I$ under the concept C, defined as:

$$v_C^I(d) = \sum_{i \in \{1, \ldots, m\}} \{w_i \mid d \in C_i^I\} \tag{2}$$

We refer the interested reader to [14,15,25] for a more precise account of the properties of the operator.

In the context of Knowledge Representation, tooth expressions provide a powerful tool to represent concepts. Tooth operators have indeed been applied in DL with a variety of goals. As already mentioned, in [14,16] a link between tooth-expressions and linear classifier has been established. In [16], in particular, it was shown that even simple tooth-expressions are expressive enough to represent complex concepts derived from real use cases in the context of the Gene Ontology.

Tooth operators are also useful in the representation of different cognitively relevant phenomena related to human concept combination and categorisation [27,28]. More precisely, the representation of tooth expressions is inspired by the design of Prototype Theory [29]. Tooth operators, and generally weighted logics [22], are thus more cognitively grounded than standard logic languages, allowing for a representation of concepts that is, arguably, more in line with the way humans think of them.

In particular, Tooth expressions are equivalent to standard DNFs, i.e., canonical normal form of logical formulas consisting of a disjunction of conjunctions of literals [16].

Let us imagine, for instance, to model the explanation for approving a loan from a loan agent, as described in the Introduction, by means of the tooth-operator. This could be captured through an axiom using a tooth expression as follows: $\exists isGranted.Loan \sqsubseteq \mathbb{W}^1((Parent, -1), (Rich, 2), (Married, 1))$.

The practical advantages for knowledge acquisition and cognitive science are thus gained without any increase in computational complexity: adding Tooth operators to $\mathcal{ALC}$ does not increase the expressivity of the language. The reason is that $\mathcal{ALC}$ is closed under Boolean operators, so any Tooth concept can be translated into a DNF of concepts of $\mathcal{ALC}$, see ([25], Sec. 3.1). Moreover, any ontology in $\mathcal{ALC}$ plus Tooth concepts can be translated into an ontology in the language of $\mathcal{ALC}$, see ([14] Sec. 2).

By representing our running example by $D = \mathbb{W}^1((A, -1), (B, 2), (C, 1))$, we show that it is extensionally equivalent to the DNF $(\neg A \sqcap C) \sqcup B$. In one direction, if $d \in ((\neg A \sqcap C) \sqcup B)^I$, then $d \in (\Delta^I \setminus A^I) \cap C^I$ or $d \in B^I$. In the first case, d scores 1 because $d \in C^I$; in the second case, d scores 2 because $d \in B^I$. Therefore, in both cases, $v_D^I(d) \geq 1$, so $d \in D^I = \mathbb{W}^1((A, -1), (B, 2), (C, 1))^I$.

In the other direction, suppose by contraposition that $d \notin ((\neg A \sqcap C) \sqcup B)^I$. So $d \notin (\Delta^I \setminus A^I) \cap C^I$ and $d \notin B^I$. We have two cases, if $d \notin (\Delta^I \setminus A^I)$, then $d \in A^I$, so d scores -1. Since $d \notin B^I$, d does not score 2, so $v_D^I(d) < 1$. If $d \notin C^I$, then d does not score 1, and since $d \notin B^I$, again $v_D^I(d) < 1$. Thus, in both cases, $v_D^I(d) < 1$, so $d \notin D^I = \mathbb{W}^1((A, -1), (B, 2), (C, 1))^I$.

2.2 Disjunctive Normal Forms - Preliminary Definitions

A disjunctive normal form (DNF) is a logical formula consisting of a disjunction of one or more conjunctions, of one or more literals. It can also be described as an OR of ANDs, as the only propositional operators in DNF are the and ($\wedge$), the or ($\vee$), and the negation ($\neg$). In our study, we used DL symbols ($\sqcap$, $\sqcup$) to interpret conjunctions and disjunctions of concepts.

Henceforth, we will follow the definitions proposed by Darwiche and Marquis [12]. Accordingly, DNF is a strict subset of the Negation Normal Form language. An NNF formula can be characterised as a *rooted, directed, acyclic* graph, where each leaf node is labeled with a propositional variable or its negation, and each internal node is labeled with a conjunction or a disjunction. A DNF is a *flat* NNF, i.e., an NNF whose maximum number of edges from the root to some leaf is 2. Moreover, DNFs satisfies the property of *simple conjunction*, i.e., each propositional variable occurs at most once in each conjunction. An example is provided in Fig. 1.

One can consider different NNF subsets by imposing one or more of the following conditions on the formulas: (i) *Decomposability*: an NNF is decomposable (DNNF) iff for each conjunction in the NNF, the conjuncts do not share variables. Each DNF is decomposable by definition. (ii) *Determinism*: an NNF is deterministic (d-NNF) iff for each disjunction in the NNF, every two disjuncts are logically contradictory. (iii) *Smoothness*: NNFs satisfy smoothness (sd-NNF) iff for each disjunction formula, each disjunct mentions the same variables. When looking at DNF, the class of formulas satisfying determinism and smoothness is called MODS.

In what follows, we will consider three sets of DNF, obtained by adding different conditions on the formulae (and leading to formulas of different sizes).

- **DNF1:** Simple (decomposable) DNFs ($DNF1 \subsetneq DNNF$), corresponding to the shorter formulas. The only requirement for the formulas is to satisfy the property of simple conjunction. See (i) in Fig. 1 for an example.
- **DNF2:** Deterministic DNFs ($DNF2 \subsetneq d - NNF$), for which each couple of disjuncts is required to be logically contradictory. See (ii) in Fig. 1 for an example.

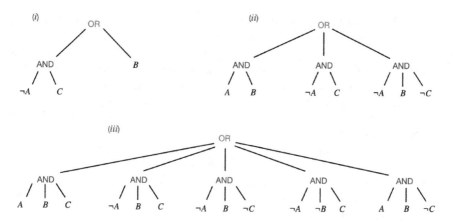

Fig. 1. Three different variants of the same DNF modeling the running example.

- **DNF3:** Deterministic, smooth DNFs ($DNF3 \subsetneq MODS$), corresponding to the longest possible DNFs. DNF3 collect all the formula models. See (*iii*) in Fig. 1 for an example.

3   Evaluating the Interpretability of Explanations

The notion of interpretability of symbolic representations has gained popularity in recent years (e.g., [1,4,20]), also due to an increasing interest in Explainable AI. How to precisely characterise interpretability is however far from being obvious, and there is, in general, no consensus on a precise definition.

From now on, we adhere to the taxonomy of interpretability evaluation proposed by Doshi-Velez and Kim [13] which supports using 'human-grounded metrics' with real users to evaluate the perceived quality of an explanation. According to this view, the evaluation focuses on the perceived interpretability of explanations rather than in their mechanistic creation. Thus, it is not important how the explanations are computed, but whether these explanations are perceived as interpretable by humans.

To operationalise this idea, different strategies have been adopted in the literature (see e.g., [1,9,20]). In order to measure the interpretability of an explanation, subjects are usually asked to perform the same task (often, a classification task) using different explanation formats. Across the different studies, the evaluation metrics can then vary, but they normally range between four metrics, namely *accuracy* (how many times did the subjects reply correctly), *time of response* (how fast they were in carrying out the task), *confidence* (how confident did they feel in their reply), and *perceived understandability* (to what extent an explanation is perceived as understandable by the user). In [20], for instance, the interpretability of decision tables, binary decision trees, and propositional rules is measured by combining the metrics of accuracy, time of response, and confidence. Allahyari and Lavesson [1] focus on the interpretability of decision tree models and rule-based models, using the perceived understandability as the only metric for the evaluation. The metrics of accuracy, time of response, confidence, and perceived understandability are taken into account in [9] to measure the interpretability of decision trees. More precisely, the paper extends Trepan [11], an algorithm that explains ANNs by means of decision trees, to include ontologies that model domain knowledge when generating explanations. The paper shows that decision trees generated taking into account domain knowledge are perceived as more understandable by users.

Booth et al. [4] compare different propositional theories and evaluates their interpretability in different domains of application. To the best of our knowledge, [4] constitutes the most thorough attempt to evaluate the interpretability of different logical languages in terms of human-grounded metrics. In their study, the authors presented subjects with natural language explanations translating different propositional languages (varying from DNFs, CNFs and other variations of NNFs), across different domains. They thus evaluated subjects' comprehension of these explanations in terms of accuracy, confidence, and time of response. They observed that while decomposability resulted in a statistically significant increase in confidence, simple conjunction did not always show an effect in their dataset. Interestingly, they also observed that the domain, in which the explanations were presented, affected the perceived understandability of the formulas.

In the following, we present a user study that compares the interpretability of tooth expressions and DNF formulas. In the study, respondents were asked to carry out two tasks using Tooth expressions and DNF formula. We evaluated the interpretability of formulas by means of accuracy in the responses, time of response, confidence in the reply, and perceived understandability of the formula used. To avoid any bias due to prior knowledge about a certain domain, we kept the presentation of the input at an abstract level, that is, respondents were provided with logical formulas not bounded to any domain in particular.

We use variables (e.g., A, B, C) for concepts occurring in the DNFs as well as for concepts occurring in the Tooth expressions. Formally, those variables range over concepts of $\mathcal{ALC}$. However, in practice, we do not present participants concepts defined by means of restricted quantifications (i.e., $\forall R.C$ and $\exists R.C$) and we focus on the Boolean operators of $\mathcal{ALC}$. Moreover, concepts in the scope of the tooth expression are simple, i.e., we do not allow for Boolean combinations. These two simplifications allow for a direct comparison between Tooth and DNFs. More complex cases shall be analysed in a longer dedicated study.

4   Experimental Evaluation

The main research hypothesis in which we were interested was whether Tooth operators are more effective and perceived as more interpretable than DNFs by human users. More precisely, we were interested in determining under what metrics this was the case (see Sect. 3), and for which types of DNF formulas (see Sect. 2.2). To verify or refute this hypothesis we designed and ran a user study.

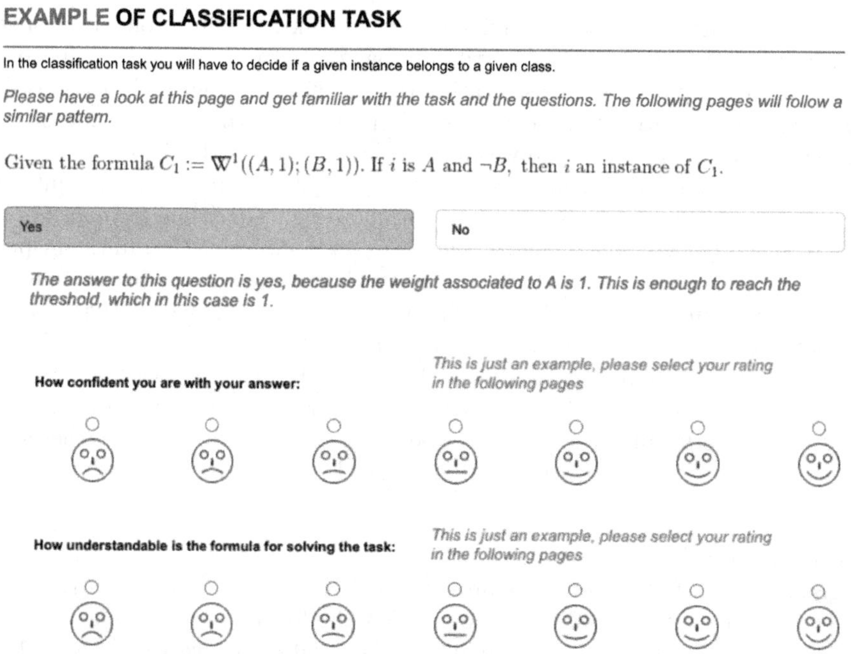

EXAMPLE OF CLASSIFICATION TASK

In the classification task you will have to decide if a given instance belongs to a given class.

Please have a look at this page and get familiar with the task and the questions. The following pages will follow a similar pattern.

Given the formula $C_1 := \mathbb{W}^1((A, 1); (B, 1))$. If i is A and $\neg B$, then i an instance of C_1.

Yes	No

The answer to this question is yes, because the weight associated to A is 1. This is enough to reach the threshold, which in this case is 1.

How confident you are with your answer:

This is just an example, please select your rating in the following pages

How understandable is the formula for solving the task:

This is just an example, please select your rating in the following pages

Fig. 2. The introductory page of the classification task for the Tooth-operator questionnaire.

4.1   Method

Materials. We used examples of concepts defined by means of DNF formulas (the three variants) and by means of the Tooth operator. We had 6 concept definitions of different complexities, varying in the number of symbols used and length. For each concept, we constructed four formulas, one for each of the formats (i.e., DNF1, DNF2, DNF3 and Tooth expression). In this manner, we obtained 24 distinct concept definitions. We had two questionnaires, one for the DNFs and one for Tooth expressions. In the user study, each participant was shown a total of twelve formulas corresponding to concept definitions. That is, participants were asked to carry out both questionnaires, in separate sessions, in random order. Concept definitions were randomly shuffled for each of the participants in the user study.

Procedure. The experiment used an online questionnaires on the usage of logical formulas to carry out certain tasks. The questionnaire was run in a controlled environment (i.e., in a classroom). The questionnaire contained an introductory and an experimental phase. In the introductory phase, subjects were shown a short description of either DNFs or Tooth operator, and how its semantics is determined. Each introduction had the same duration, and consisted of the same number of slides (and examples) for DNFs and Tooth expressions.

The experimental phase was divided into two tasks: classification, and inspection. Each task starts with an instruction page describing the task to be performed (an example for the classification task is shown in Fig. 2). In these tasks the participants were presented with six formulas corresponding to one of the two representations (one of the variants of the DNFs and the Tooth operator). In the classification task, subjects were asked to decide if a certain combination of literals is an instance of a given formula (e.g., *Given the formula* $C_1 :=$ $(\neg A \sqcap C) \sqcup B$. *If i is* $\neg A$, *B, and* $\neg C$, *then i is an instance of* C_1). In the inspection task, participants had to decide on the truth value of a particular statement, referring to if some given conditions of an instance are necessary for the instance to belong to a given class (e.g., *Given the formula* $C_1 := (\neg A \sqcap C) \sqcup B$. *Having B is necessary for being classified as* C_1). The main difference between the two types of questions used in the two tasks is that the former provides all details necessary for performing the decision, whereas the latter only specifies whether a subset of the features influence the decision. In these two tasks, for each formula, we recorded:

- Correctness of the response.
- Confidence in the response, as provided on a Likert scale from 1 to 7.
- Response time measured from the moment the formula was presented.
- Perceived formula understandability, as provided on a Likert scale from 1 to 7.

Participants. 58 participants volunteered to take part in the experiment. The participants were recruited among students with different backgrounds. In particular we had two groups of students, 33 students with a background in computer science and 25 students with a background in philosophy. Each group repeated the questionnaire twice, once using DNFs and once using Tooth expressions. In the analysis, we will denote these groups as GroupI and GroupII respectively.

4.2   Results

As it can be appreciated in Table 1, when looking at the two groups together, respondents carried out both tasks correctly, performing better in the classification task than in the inspection task. This is in line with our assumption that the classification task was simpler than the inspection task, due to the fact that more information was provided for making the decision. Remarkably, the influence of the type of formula on the percentage of correct answers is not significant in our dataset. More specifically, the answers to tasks containing DNFs are slightly more accurate than those containing Tooth expressions, but this difference is not statistically significant. Nonetheless, we observed a significant influence ($p < .0001$) of Tooth expressions on the time of response within both tasks, showing that when using Tooth operators respondents carried out the tasks in a quicker way. This suggests that Tooth expressions are more cognitively friendly than standard DNFs. Interestingly, Tooth operators were perceived as more understandable in carrying out the inspection task. Similarly, users were more confident with their

Table 1. Mean values of correct answers, time of response, user confidence, and user understandability for formulas represented using DNFs and Tooth operator (standard deviations are reported in parenthesis).

Task	Measure	DNFs	Tooth
Classification	%Correct Responses	0.91 (0.28)	0.90 (0.29)
	Time (sec)	46.78 (58.90)	29.87 (20.72)
	Confidence	5.74 (1.32)	5.65 (1.51)
	User Understandability	5.80 (1.24)	5.55 (1.44)
Inspection	%Correct Responses	0.87 (0.34)	0.83 (0.37)
	Time (sec)	28.67 (28.78)	19.78 (19.78)
	Confidence	5.70 (1.32)	5.82 (1.49)
	User Understandability	5.79 (1.24)	5.81 (1.43)

Table 2. Mean values of correct answers, time of response, user confidence, and user understandability for formulas represented using DNFs and Tooth operator for **GroupI** and **GroupII** (standard deviations are reported in parenthesis).

Group	Measure	DNFs	Tooth
Computer Science	%Correct Responses	0.90 (0.32)	0.88 (0.31)
	Time (sec)	37.29 (55.29)	25.23 (17.96)
	Confidence	5.98 (1.29)	5.73 (1.71)
	User Understandability	6.11 (1.17)	5.61 (1.65)
Philosophy	%Correct Responses	0.86 (0.30)	0.90 (0.34)
	Time (sec)	36.39 (28.06)	24.80 (16.77)
	Confidence	5.44 (1.28)	5.88 (1.15)
	User Understandability	5.43 (1.20)	5.84 (1.10)

answers when using Tooth operators in the inspection task. This is in line with our assumption that Tooth operators could be perceived as simpler representations when the task can benefit from a more compact representation of the concepts. On the contrary, DNFs were perceived better than Tooth operators in the classification task, and respondents were more confident with their answers.

When looking at the two groups separately (Table 2), the percentages of correct answers are slightly different when using DNFs and Tooth operators, but this difference is again not significant. Thus, generally, we can conclude that the type of formula used does not have any significant effects or interactions on the accuracy of responses. Tooth operators yielded faster responses in both groups. This seems to suggest that having more compact information, like in the case of Tooth operators, could speed up the human decision-making process.

Table 3. Mean values of correct answers, time of response, user confidence, and user understandability for formulas represented using DNF1, DNF2, DNF3 and Tooth operator for both groups (standard deviations are reported in parenthesis).

Measure	DNF1	DNF2	DNF3	Tooth
%Correct Responses	0.91 (0.28)	0.90 (0.30)	0.78 (0.42)	0.83 (0.38)
Time (sec)	21.03 (10.84)	25.01 (19.99)	39.97 (42.28)	19.78 (12.81)
Confidence	6.21 (1.19)	5.72 (1.20)	5.18 (1.37)	5.82 (1.49)
User Understandability	6.14 (1.00)	5.99 (1.19)	5.24 (1.34)	5.81 (1.43)

Interestingly, faster decision making can yield more correct responses, but surprisingly faster decision-making is not always associated with highest perceived understandability and highest confidence. Respondents with computer science background were more confident with DNFs and perceived them as more understandable than Tooth operators. On the contrary, respondents with a background in philosophy found Tooth operators more understandable and were more confident with their answers when using Tooth operators. This behaviour can be motivated by the fact that computer scientists were introduced to logic and DNF formulas in their curricula, but not to Tooth operators. Thus, being more proficient in DNFs, they did not face the 'learning curve' in understanding a new representation formalism such Tooth operators. Respondents with a background in philosophy, on the other hand, studied neither DNFs nor Tooth operators. From this study, we can conclude that Tooth operators are better representation for users who are not familiar with logic, and with DNFs in particular.

When looking at results of different DNFs vs Tooth operator (Table 3), we can observe that simpler DNF formats, namely DNF1 and DNF2, yielded more accurate responses. Tooth operators perform better compared to DNF3. This is expected since formulas in DFN3 format tend to be very long (see examples in Sect. 2.2). DNF1 and DNF2 performs similarly in our study. This is expected, since they are quite similar in lengths and they do not impose a cognitive burden on the users w.r.t. DNF3 (as also shown in the previous study comparing them directly [4]). As far as time is concerned, we still observe that Tooth operators are faster than any of the DNF formats. Remarkably, the response time obtained using DNF1 is similar to the one obtained when using the Tooth operator. This can be motivated by observing that DNF1 format can be considered still a concise representation. Thus, the 'interformat' analysis seems to suggest that DNF1 and Tooth operator have quite similar understandability from the performance point of view and also from the subjective point of view. On the other hand, DFN2 and DNF3 require longer time of response and were perceived as less understandable than Tooth operators.

5   Conclusion and Future Works

In this paper, we studied the intepretability of threshold operators, by comparing them with a standard logical formalism, i.e. the DNFs. To model threshold operators in a logical setting and to facilitate the comparison with DNFs, we presented the threshold operators as concept constructors on top of $\mathcal{ALC}$, i.e. the Tooth expressions. Then, we proposed a user study aiming at comparing the interpretability of Tooth expressions and DNFs.

On the one hand, DNFs are conventionally considered a benchmark in terms of both expressivity and interpretability of logical languages [12]. On the other hand, Tooth expressions [25] provide a more concise representation of formulas. Furthermore, they are cognitively grounded, since their design is inspired by Prototype Theory [29]. Thus, they should allow for a representation of concepts that is, arguably, more in line with the way humans think of them. We hypothesised tooth expressions to be generally more interpretable than DNFs.

In the user study, we compared Tooth expressions with equivalent DNFs of different complexity and length, by imposing different conditions on the DNFs used (see Appendix A). We asked users to carry out two distinct tasks, namely classification and inspection (see Sect. 4), using Tooth expressions and DNFs. The interpretability of Tooth expressions and DNFs was measured through human-grounded metrics, namely accuracy in the responses, time of response, confidence in the responses, and perceived understandability.

In line with our hypothesis, the study revealed that Tooth expressions are generally faster to use, leading to a lower time of response. This was observed across all different DNFs formats considered in the study. Moreover, Tooth expressions were perceived as more understandable than DNFs in the inspection task (suggesting that they are better suited to tasks that benefit from a more compact representation of knowledge). The same was not generally observed in the classification task. Whilst the time of response was much lower for Tooth expressions than DNFs and the percentage of correct responses was almost the same for Tooth expressions and DNFs, the confidence in the reply and the perceived understandability were higher in the case of DNF formulas. By distinguishing different DNF formats, we observed that longer DNFs (e.g., DNF3) were perceived as less understandable than Tooth expressions. This result was also affected by the background of the respondents. Tooth operators, in particular, resulted in better performances and in a higher level of perceived understandability for users who were not familiar with logic.

The results obtained open several directions for future work. Firstly, we plan a second user study, where both Tooth expressions and DNFs are translated into natural language. This would allow to further test whether the algorithm of classification which stands behind the Tooth operator is more interpretable and easy to use. Secondly, we plan to compare decision trees and Tooth expressions [7]. Decision trees and Tooth expressions seem to have complementary pros and cons when considered in the context of XAI. Analysing the different performances of users in either the representations might provide useful insights on which representation format would be more suitable in relation to different contexts, tasks, and applications.

Acknowledgment. The authors thank Oliver Kutz, Nicolas Troquard, Pietro Galliani, and Antonella De Angeli for taking the pre-test and providing precious feedback about the user study.

A   Examples used in the questionnaires

1. – DNF1: $A \sqcup B$
    – DNF2: $A \sqcup (\neg A \sqcap B)$
    – DNF3: $(A \sqcap B) \sqcup (\neg A \sqcap B) \sqcup (A \sqcap \neg B)$
    – Tooth: $\mathbb{W}^1((A, 1), (B, 1))$

2. – DNF1: $(\neg A \sqcap C) \sqcup B$
    – DNF2: $(A \sqcap B) \sqcup (\neg A \sqcap C) \sqcup (\neg A \sqcap B \sqcap \neg C)$
    – DNF3: $(A \sqcap B \sqcap C) \sqcup (\neg A \sqcap B \sqcap C) \sqcup (\neg A \sqcap B \sqcap \neg C) \sqcup (\neg A \sqcap \neg B \sqcap C) \sqcup (A \sqcap B \sqcap \neg C)$
    – Tooth: $\mathbb{W}^2((\neg A, 1), (B, 2), (C, 1)) \equiv \mathbb{W}^1((A, -1), (B, 2), (C, 1))$

3. – DNF1: $(\neg A \sqcap B) \sqcup C$
    – DNF2: $(\neg A \sqcap B) \sqcup (A \sqcap \neg B \sqcap C) \sqcup (A \sqcap B \sqcap C) \sqcup (\neg A \sqcap \neg B \sqcap C)$
    – DNF3: $(\neg A \sqcap B \sqcap C) \sqcup (\neg A \sqcap B \sqcap \neg C) \sqcup (A \sqcap \neg B \sqcap C) \sqcup (A \sqcap B \sqcap C) \sqcup (\neg A \sqcap \neg B \sqcap C)$
    – Tooth: $\mathbb{W}^2((A, -1), (B, 2), (C, 3))$

4. – DNF1: $(A \sqcap B) \sqcup (B \sqcap C) \sqcup (A \sqcap C)$
    – DNF2: $(A \sqcap B) \sqcup (A \sqcap \neg B \sqcap C) \sqcup (\neg A \sqcap B \sqcap C)$
    – DNF3: $(A \sqcap B \sqcap C) \sqcup (\neg A \sqcap B \sqcap C) \sqcup (A \sqcap \neg B \sqcap C) \sqcup (A \sqcap B \sqcap \neg C)$
    – Tooth: $\mathbb{W}^2((A, 1), (B, 1), (C, 1))$

5. – DNF1: $(A \sqcap D) \sqcup (A \sqcap B \sqcap C) \sqcup (D \sqcap B) \sqcup (D \sqcap C)$
    – DNF2: $(A \sqcap D) \sqcup (A \sqcap B \sqcap C \sqcap \neg D) \sqcup (\neg A \sqcap B \sqcap D) \sqcup (\neg A \sqcap \neg B \sqcap C \sqcap D)$
    – DNF3: $(\neg A \sqcap \neg B \sqcap C \sqcap D) \sqcup (\neg A \sqcap B \sqcap \neg C \sqcap D) \sqcup (\neg A \sqcap B \sqcap C \sqcap D) \sqcup (A \sqcap \neg B \sqcap \neg C \sqcap D) \sqcup (A \sqcap \neg B \sqcap C \sqcap D) \sqcup (A \sqcap B \sqcap \neg C \sqcap D) \sqcup (A \sqcap B \sqcap C \sqcap \neg D) \sqcup (A \sqcap B \sqcap C \sqcap D)$
    – Tooth: $\mathbb{W}^5((A, 3), (B, 1), (C, 1), (D, 4))$

6. – DNF1: $(A \sqcap B) \sqcup (A \sqcap C) \sqcup (A \sqcap D) \sqcup (B \sqcap D)$
    – DNF2: $(A \sqcap B \sqcap \neg D) \sqcup (\neg A \sqcap B \sqcap C \sqcap D) \sqcup (A \sqcap \neg B \sqcap C \sqcap \neg D) \sqcup (\neg A \sqcap B \sqcap \neg C \sqcap D) \sqcup (A \sqcap D)$
    – DNF3: $(\neg A \sqcap B \sqcap \neg C \sqcap D) \sqcup (\neg A \sqcap B \sqcap C \sqcap D) \sqcup (A \sqcap \neg B \sqcap \neg C \sqcap D) \sqcup (A \sqcap \neg B \sqcap C \sqcap \neg D) \sqcup (A \sqcap \neg B \sqcap C \sqcap D) \sqcup (A \sqcap B \sqcap \neg C \sqcap \neg D) \sqcup (A \sqcap B \sqcap \neg C \sqcap D) \sqcup (A \sqcap B \sqcap C \sqcap \neg D) \sqcup (A \sqcap B \sqcap C \sqcap D)$
    – Tooth: $\mathbb{W}^3((A, 2), (B, 1.5), (C, 1), (D, 1.5))$

7. – DNF 1: $(A \sqcap B) \sqcup (A \sqcap C \sqcap D) \sqcup (B \sqcap C \sqcap D)$
    – DNF 2: $(A \sqcap B) \sqcup (\neg A \sqcap B \sqcap C \sqcap D) \sqcup (A \sqcap \neg B \sqcap C \sqcap D)$
    – DNF 3: $(A \sqcap B \sqcap C \sqcap D) \sqcup (A \sqcap B \sqcap \neg C \sqcap \neg D) \sqcup (A \sqcap B \sqcap \neg C \sqcap D) \sqcup (A \sqcap B \sqcap C \sqcap \neg D) \sqcup (\neg A \sqcap B \sqcap C \sqcap D) \sqcup (A \sqcap \neg B \sqcap C \sqcap D)$
    – Tooth: $\mathbb{W}^4((A, 2), (B, 2), (C, 1), (D, 1))$

References

1. Allahyari, H., Lavesson, N.: User-oriented assessment of classification model understandability. In: SCAI 2011 Proceedings, vol. 227, pp. 11–19. IOS Press (2011)
2. Barredo Arrieta, A., et al.: Explainable Explainable Artificial Intelligence (XAI): concepts, taxonomies, opportunities and challenges toward responsible AI. Inf. Fusion **58**(October 2019), 82–115 (2020). https://doi.org/10.1016/j.inffus.2019.12.012
3. Bishop, C.M.: Pattern Recognition and Machine Learning, 5th Edition. Information science and statistics, Springer, New York (2007). ISBN 9780387310732. https://www.worldcat.org/oclc/71008143
4. Booth, S., Muise, C., Shah, J.: Evaluating the interpretability of the knowledge compilation map. In: Kraus, S. (ed.) Proceedings of IJCAI, pp. 5801–5807 (2019)
5. Coba, L., Confalonieri, R., Zanker, M.: RecoXplainer: a library for development and offline evaluation of explainable recommender systems. IEEE Comput. Intell. Mag. **17**(1), 46–58 (2022). https://doi.org/10.1109/MCI.2021.3129958
6. Confalonieri, R., Coba, L., Wagner, B., Besold, T.R.: A historical perspective of explainable artificial intelligence. WIREs Data Min. Knowl. Disc. **11**(1), e1391 (2021). https://doi.org/10.1002/widm.1391
7. Confalonieri, R., Galliani, P., Kutz, O., Porello, D., Righetti, G., Troquard, N.: Towards knowledge-driven distillation and explanation of black-box models. In: Confalonieri, R., Kutz, O., Calvanese, D. (eds.) Proceedings of the Workshop on Data meets Applied Ontologies in Explainable AI (DAO-XAI 2021) part of Bratislava Knowledge September (BAKS 2021), CEUR Workshop Proceedings, Bratislava, Slovakia, 18–19 September 2021, vol. 2998. CEUR-WS.org (2021)
8. Confalonieri, R., Lucchesi, F., Maffei, G., Solarz, S.C.: A unified framework for managing sex and gender bias in AI models for healthcare. In: Sex and Gender Bias in Technology and Artificial Intelligence. Elsevier, pp. 179–204 (2022)
9. Confalonieri, R., Weyde, T., Besold, T.R., Moscoso del Prado Martín, F.: Using ontologies to enhance human understandability of global post-hoc explanations of black-box models. Artif. Intell. **296** (2021). https://doi.org/10.1016/j.artint.2021.103471
10. Confalonieri, R., Weyde, T., Besold, T.R., del Prado Martín, F.M.: Trepan reloaded: a knowledge-driven approach to explaining black-box models. In: Proceedings of the 24th European Conference on Artificial Intelligence, pp. 2457–2464 (2020). https://doi.org/10.3233/FAIA200378
11. Craven, M.W., Shavlik, J.W.: Extracting tree-structured representations of trained networks. In: NIPS 1995, pp. 24–30. MIT Press (1995)
12. Darwiche, A., Marquis, P.: A knowledge compilation map. J. Artif. Intell. Res. **17**, 229–264 (2002). https://doi.org/10.1613/jair.989
13. Doshi-Velez, F., Kim, B.: Towards a rigorous science of interpretable machine learning (2017)
14. Galliani, P., Kutz, O., Porello, D., Righetti, G., Troquard, N.: On knowledge dependence in weighted description logic. In: Calvanese, D., Iocchi, L. (eds.) GCAI 2019. Proceedings of the 5th Global Conference on Artificial Intelligence, EPiC Series in Computing, Bozen/Bolzano, Italy, 17–19 September 2019, vol. 65, pp. 68–80. EasyChair (2019)

15. Galliani, P., Kutz, O., Troquard, N.: Perceptron operators that count. In: Homola, M., Ryzhikov, V., Schmidt, R.A. (eds.) Proceedings of the 34th International Workshop on Description Logics (DL 2021) part of Bratislava Knowledge September (BAKS 2021), CEUR Workshop Proceedings, Bratislava, Slovakia, 19–22 September 2021, vol. 2954. CEUR-WS.org (2021)

16. Galliani, P., Righetti, G., Kutz, O., Porello, D., Troquard, N.: Perceptron connectives in knowledge representation. In: Keet, C.M., Dumontier, M. (eds.) EKAW 2020. LNCS (LNAI), vol. 12387, pp. 183–193. Springer, Cham (2020). https://doi.org/10.1007/978-3-030-61244-3_13

17. Garcez, A.D., Gori, M., Lamb, L.C., Serafini, L., Spranger, M., Tran, S.N.: Neural-symbolic computing: an effective methodology for principled integration of machine learning and reasoning. IfCoLoG J. Log. Appl. **6**(4), 611–631 (2019)

18. Guidotti, R., Monreale, A., Ruggieri, S., Turini, F., Giannotti, F., Pedreschi, D.: A survey of methods for explaining black box models. ACM Comp. Surv. **51**(5), 1–42 (2018)

19. Hind, M.: Explaining explainable AI. XRDS **25**(3), 16–19 (2019). https://doi.org/10.1145/3313096

20. Huysmans, J., Dejaeger, K., Mues, C., Vanthienen, J., Baesens, B.: An empirical evaluation of the comprehensibility of decision table, tree and rule based predictive models. Decis. Support Syst. **51**(1), 141–154 (2011)

21. Mariotti, E., Alonso, J.M., Confalonieri, R.: A framework for analyzing fairness, accountability, transparency and ethics: a use-case in banking services. In: 30th IEEE International Conference on Fuzzy Systems, FUZZ-IEEE 2021, Luxembourg, 11–14 July 2021, pp. 1–6. IEEE (2021). https://doi.org/10.1109/FUZZ45933.2021.9494481

22. Masolo, C., Porello, D.: Representing concepts by weighted formulas. In: Borgo, S., Hitzler, P., Kutz, O. (eds.) Formal Ontology in Information Systems - Proceedings of the 10th International Conference, FOIS 2018, Frontiers in Artificial Intelligence and Applications, Cape Town, South Africa, 19–21 September 2018, vol. 306, pp. 55–68. IOS Press (2018). https://doi.org/10.3233/978-1-61499-910-2-55

23. Mittelstadt, B., Russell, C., Wachter, S.: Explaining explanations in AI. In: Proceedings of the Conference on Fairness, Accountability, and Transparency - FAT* 2019, pp. 279–288. ACM Press, New York (2019). https://doi.org/10.1145/3287560.3287574

24. Parliament and Council of the European Union: General Data Protection Regulation (2016)

25. Porello, D., Kutz, O., Righetti, G., Troquard, N., Galliani, P., Masolo, C.: A toothful of concepts: towards a theory of weighted concept combination. In: Simkus, M., Weddell, G.E. (eds.) Proceedings of the 32nd International Workshop on Description Logics, CEUR Workshop Proceedings, Oslo, Norway, 18–21 June 2019, vol. 2373. CEUR-WS.org (2019)

26. Ribeiro, M.T., Singh, S., Guestrin, C.: "Why should I trust you?": explaining the predictions of any classifier. In: Proceedings of the 22nd International Conference on Knowledge Discovery and Data Mining, KDD 2016, pp. 1135–1144. ACM (2016)

27. Righetti, G., Masolo, C., Troquard, N., Kutz, O., Porello, D.: Concept combination in weighted logic. In: Sanfilippo, E.M., et al. (eds.) Proceedings of the Joint Ontology Workshops 2021 Episode VII, CEUR Workshop Proceedings, vol. 2969. CEUR-WS.org (2021)

28. Righetti, G., Porello, D., Kutz, O., Troquard, N., Masolo, C.: Pink panthers and toothless tigers: three problems in classification. In: Cangelosi, A., Lieto, A. (eds.)

Proceedings of the 7th International Workshop on Artificial Intelligence and Cognition, CEUR Workshop Proceedings, vol. 2483, pp. 39–53. CEUR-WS.org (2019)

29. Rosch, E., Lloyd, B.B.: Cognition and categorization (1978)

30. Vollmer, H.: Introduction to Circuit Complexity: A Uniform Approach. Springer, Heidelberg (1999). https://doi.org/10.1007/978-3-662-03927-4

EBOCA: Evidences for BiOmedical Concepts Association Ontology

Andrea Álvarez Pérez[1], Ana Iglesias-Molina[2], Lucía Prieto Santamaría[1], María Poveda-Villalón[2], Carlos Badenes-Olmedo[2], and Alejandro Rodríguez-González[1]($\boxtimes$)

[1] Centro de Tecnología Biomédica, Universidad Politécnica de Madrid, Madrid, Spain
andrea.alvarezp@alumnos.upm.es,
{lucia.prieto.santamaria,alejandro.rg}@upm.es
[2] Ontology Engineering Group, Universidad Politécnica de Madrid, Madrid, Spain
{ana.iglesiasm,m.poveda,carlos.badenes}@upm.es

Abstract. There is a large number of online documents data sources available nowadays. The lack of structure and the differences between formats are the main difficulties to automatically extract information from them, which also has a negative impact on its use and reuse. In the biomedical domain, the DISNET platform emerged to provide researchers with a resource to obtain information in the scope of human disease networks by means of large-scale heterogeneous sources. Specifically in this domain, it is critical to offer not only the information extracted from different sources, but also the evidence that supports it. This paper proposes EBOCA, an ontology that describes (i) biomedical domain concepts and associations between them, and (ii) evidences supporting these associations; with the objective of providing an schema to improve the publication and description of evidences and biomedical associations in this domain. The ontology has been successfully evaluated to ensure there are no errors, modelling pitfalls and that it meets the previously defined functional requirements. Test data coming from a subset of DISNET and automatic association extractions from texts has been transformed according to the proposed ontology to create a Knowledge Graph that can be used in real scenarios, and which has also been used for the evaluation of the presented ontology.

Keywords: Ontology · Biomedicine · Evidences · Provenance · Semantic

1 Introduction

The availability of biomedical data has increased in recent decades [6]. This type of content, whether structured (i.e. relational databases) or unstructured (i.e. text), is usually organized in separate isles owned by companies or institutions, sometimes with proprietary formats. This heterogeneity makes it difficult to extract knowledge through them. The search for drugs, for instance, that could

O. Corcho et al. (Eds.): EKAW 2022, LNAI 13514, pp. 152–166, 2022.
https://doi.org/10.1007/978-3-031-17105-5_11

interact with a certain drug, e.g. Plaquenil, during the treatment of COVID-19 based on the experiments published in scientific publications becomes a challenging task. Articles do not usually mention the trade name of drugs, but the active ingredient, Hydroxychloroquine in this case for Plaquenil. The need for massive semantic integration of such information and the establishment of standards, as well as the inclusion of its explicit provenance, is becoming increasingly noticeable. Among the many utilities that they have, bioinformaticians have seen ontologies as a way to manage this explosion of data, facilitating both manual and automatic computer handling. Thus, in the last decades, there has been a considerable adoption of ontologies to model biomedical knowledge [5].

Several efforts have been done to integrate this biomedical knowledge in a unique and shared space of representation [4, 21]. One of the most recent works is DISNET[1], which provides researches with a platform that enables the creation of complex multilayered graphs [23] following the concepts of Human Disease Networks (HDNs) [18]. This system aims to give better insights into disease understanding [34, 39] and generate new drug repurposing hypotheses [32, 33], by putting together in an accessible knowledge base heterogeneous information that includes large-scale biomedical data obtained and integrated from both structured and unstructured sources. The information in DISNET relational database is organized in three topological levels: (i) the phenotypical (for diseases and their associated symptoms), (ii) the biological (for molecular-shifted data related to diseases including genes, proteins, metabolic pathways, genetic variants, non-coding RNAs, etc.) and (iii) the pharmacological (for drugs, their interactions and their connections to diseases).

Despite the efforts of building a unifying and complete biomedical resource, these resources sometimes lack traceability. Users using those resources need to know where each piece of information comes from, and what evidence is supporting it. Thus, completing the resources with this metadata can greatly improve decision making, which is particularly important in the biomedical domain.

In the current work, we present EBOCA, an ontology that aims to model Evidences for BiOmedical Concepts Association. It describes in two modules (i) biomedical concepts and their associations and (ii) the evidences supporting the associations with metadata and provenance. With this ontology we conceptualize the model that will allow to create a complete semantic resource based on the DISNET biomedical knowledge and associations extracted from texts and other sources enriched with metadata about the evidences supporting them.

The rest of the paper is organized as follows: Sect. 2 describes the previous works that have been carried out in the context of the ontology. Section 3 explains the methodology that has been followed to develop the ontology. Section 4 describes the ontology and its modules in detail, while Sect. 5 develops the evaluation of the ontology. Finally, Sect. 6 draws the conclusions.

---

[1] https://disnet.ctb.upm.es/.

2   Related Work

We divide this section to present similar works related to each of the two modules of EBOCA. That is, we first describe the works that have been previously carried out in the context of biomedical ontologies and secondly include the main ones related to handling evidences, metadata and provenance.

On the one hand, the ontologies that have been proposed to tackle the integration of the biomedical knowledge are multiple, diverse, and different depending on their specific context. Generally, the purpose of biomedical ontologies is to study classes of entities which are of biomedical significance in order to enable sharing complex biological information and the integration of heterogeneous databases [6]. Open Biomedical Ontologies (OBO) [21] was created as an ontology information resource, which now comprises more than 60 ontologies, accessible through BioPortal[2]. Between the different resources that it offers, one can find for instance the Human Phenotype Ontology (HPO) [22], the Sequence Ontology (SO) [14], or the PRotein Ontology (PRO) [27]. OBO Foundry is supported by the National Center for Biomedical Ontology (NCBO) and has a version that is being developed by the National Cancer Institute thesaurus (NCIt)[3]. The NCIt provides definitions, synonyms, and other information on nearly 10,000 cancers and related diseases, 17,000 single agents and related substances, and other topics related to cancer and biomedical research. Also maintained by the USA National Institutes of Health (NIH), in particular, by the National Library of Medicine (NLM), the Unified Medical Language System (UMLS)[4] is a repository that brings together a great number of health and biomedical vocabularies to enable interoperability [4]. In this line, Bio2RDF[5] is a system that uses Semantic Web technologies to build and provide a network of Linked Data for the life sciences [3]. Regarding the concept of diseases, one of the most renowned efforts to model such an entity resulted in the Disease Ontology (DO)[6], which has been developed as a standard providing descriptions of human disease terms, phenotype characteristics and related medical vocabulary disease concepts [37]. In the specific case of rare diseases, Orphanet[7] developed Orphanet Rare Disease Ontology (ORDO) [40]. Moreover, other resources have centered their scope in other biomedical entities. DisGeNET integrates associations between genes and variants to human diseases [29]. In the context of associations between biomedical entities, the Semanticscience Integrated Ontology (SIO) provides a simple, integrated ontology of types and relations for rich description of objects, processes and their attributes [13]. As for proteins, other databases and terminologies include UniProt [36] or NeXtProt [42]. WikiPathways represents and integrated data regarding biological pathways [25]. Finally, drugs and their related information (e.g. their interactions) have been modeled and integrated in several

---

[2] https://bioportal.bioontology.org/.

[3] https://ncithesaurus.nci.nih.gov/.

[4] https://uts.nlm.nih.gov/uts/.

[5] https://bio2rdf.org/.

[6] https://disease-ontology.org/.

[7] https://www.orpha.net/.

sources. Between those, we can find ChEMBL [26], the Comparative Toxicoge-nomics Database (CTD) [11], DrugBank [41] or TWOSIDES[8].

On the other hand, not so many ontologies include evidence information. With the exponential adoption of computational techniques able to discover new knowl-edge, the need to track the evidence and provenance of these techniques is becom-ing increasingly important. The Evidence and Conclusion Ontology (ECO) [17] was developed to tackle this issue. It provides a controlled vocabulary that enables describing the type of evidence of an assertion, with a focus on the biomedical domain. This ontology is maintained and curated with active participation from the community, as it is currently used by projects such as the Gene Ontology (GO) [9] and DisGeNET [29]. To track evidences, more metadata is required, and many ontologies have been developed for this purpose and are widely used. Some examples are the PROV Ontology (PROV-O) [24], its extension Provenance, Authoring and Versioning (PAV) [8], and DCMI Metadata Terms[9]. The module EBOCA Evidences reuses entities from these well-established ontologies to provide provenance and metadata of biomedical associations evidences, that may come from both curated data sources or inferred from unstructured texts.

3   Methodology

The EBOCA Ontology was developed following the guidelines provided by the Linked Open Terms (LOT) methodology [31]. LOT, based on the NeOn method-ology [38], is a lightweight methodology for ontology and vocabulary develop-ment. It includes four major stages: Requirements Specification, Implementation, Publication, and Maintenance (Fig. 1). In this section, we describe these stages and how they have been applied and adapted to the development of EBOCA.

Requirements Specification. This first stage involves the activities that lead to defining the requirements that the ontology must meet. Those are the purpose and scope of the ontology, i.e., the objective of the ontology and its extent, the knowledge it models; and the requirements specified in the form of competency questions and/or affirmative statements. Both purpose and scope are specified in the ontology documentation. The requirements can be found in the repositories of each module, and accessed through the ontology portal[10].

Implementation. The goal of this stage is to build the ontology using a formal language, using the requirements identified in the previous stage as guidance. The first version of the ontology is conceptualized based on these requirements, and subsequently refined by verifying the model with domain experts. The con-ceptualization is carried out representing the ontology graphically following the Chowlk notation [7] and implemented in OWL 2 with Protégé. The evaluation of the ontology is carried out using (i) SPARQL queries and Themis [15] to vali-date the requirements, (ii) OOPS! [30] to identify modelling pitfalls and (iii) the Pellet and HermiT reasoners to check for inconsistencies.

---

[8] https://nsides.io/.

[9] https://www.dublincore.org/specifications/dublin-core/dcmi-terms/.

[10] https://w3id.org/eboca/portal.

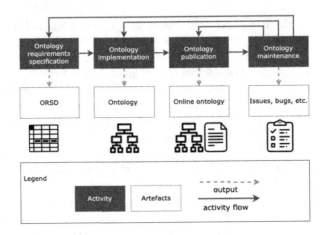

Fig. 1. LOT methodology [31]

Publication. The publication stage refers to the activities carried out to make the ontology and its documentation available. The documentation is generated with Widoco [16], and it is published with a W3ID URI[11]. The published EBOCA resources can be accessed through the ontology portal.

Maintenance. Finally, the maintenance stage aims to ensure that the ontology is updated with error corrections and new requirements. The EBOCA ontology enables the gathering of new requirements and issues through GitHub and GitLab repositories.

4   Ontology Description

The EBOCA Ontology describes the evidences of associations between biomedical concepts. It is composed of two parts, one for modelling biomedical concepts and associations, EBOCA SEM-DISNET; and the other for representing the evidences of these associations with metadata and provenance information, EBOCA Evidences. They are designed and developed as separated modules with one class of connection, `sio:SIO_000897` ("Association"), that will be described further in this section. All related resources of this ontology are publicly available online[12].

4.1   EBOCA SEM-DISNET

The EBOCA SEM-DISNET module[13] is designed to represent the associations of common biomedical concepts. These concepts include principally: diseases, phenotypes, genes, genetical variants, biological pathways, drugs, proteins and targets. The associations link pairs of concepts, e.g. gene-disease or drug-disease

---

[11] https://w3id.org/eboca/sem-disnet, https://w3id.org/eboca/evidences.

[12] https://w3id.org/eboca/portal.

[13] https://w3id.org/eboca/sem-disnet.

association. This module is also focused on representing the data contained in DISNET relational database [23] and give them a semantic structure. These data are organized in three layers: (i) the phenotypical (in which disease-phenotype relationships are gathered from applying text mining processes to Wikipedia, PubMed and MayoClinic), the biological (with diseases and their relationships to genes, proteins, genetical variants, pathways and so on, extracted from structured sources as DisGeNET) and the pharmacological (which stores drugs and drug-related information from structured sources as ChEMBL).

The requirements of this module are formalized as 15 competency questions, validated through SPARQL queries (see Sect. 5). These functional requirements make reference to the scope of the ontology, which has been defined as the already mentioned biomedical concepts as well as the association between those concepts. The final intended users of EBOCA SEM-DISNET could mainly be research scientists in the field of life science and computer science, and some of the most direct uses would be the aforementioned generation of drug repurposing hypotheses or the obtainment of a better understanding of diseases.

This module of EBOCA is mostly built from the reuse of a wide amount of existing ontologies. That is, the majority of its entities have been obtained from previously developed terminologies, when possible. The creation of new concepts from scratch has only been implemented when it was totally unavoidable. The general design pattern used to model this module is based on DisGeNET [35], that provides RDF resources integrating information of associations between genes and diseases (and others such as variants and diseases). This design pattern was suitable enough to represent other types of associations as well. We considered that the details and representations established by this patterns were in accordance with the information to be modeled from the other named associaditons. EBOCA SEM-DISNET consists of a total of 29 classes and 33 object properties reused from different previously published vocabularies. However, in some occasions, it was not accurate to reuse certain classes due to their absence or because, even if they existed, they did not fit the specifications of concept to be modeled. This was the case of 5 classes and 12 object properties, which have been specifically created for EBOCA SEM-DISNET.

On the one hand, for both the classes modeling concepts and associations, most terms have been reused from NCIt, closely followed by SIO, and to a lesser extent by ChEMBL and ORDO. Furthermore, the classes corresponding to associations between concepts were incorporated into a class hierarchy, and was represented independently of the main diagram for the sake of clarity. In this hierarchy, all association entities were modeled as subclasses of the `sio:SIO_000897` class, denoting "Association" and chosen to link both EBOCA modules. It can also occur that the reused class is not directly related to the "Association" class. In these cases, for example, for the `cco:Mechanism` class, the strategy consisted in modelling a SEM-DISNET class named `DrugTargetAssociation`, which is a subclass of ChEMBL's `cco:Mechanism` and "Association".

On the other hand, with respect to properties, it was intended that the relationships between classes were as homogeneous as possible. However, while most have been reused from SIO (almost half of them), it has been necessary to reuse some from ChEMBL and the NCIt. SEM-DISNET relationships were created with the purpose of joining the meaning of some of the reused object properties with `sio:SIO_000628` (which means 'refers to') or `sio:SIO_000212` ('is referred to by'). We followed the same strategy as for the classes. An object property was modeled under the SEM-DISNET namespace and classified as a subproperty of the mentioned 'refers to' or 'is referred to by' according to its meaning, and the one from the original ontology, establishing a hierarchy[14]. For example, we created the object property `drugForMechanism` which is at the same time subproperty of `cco:hasMechanism` and `sio:SIO_000212`.

EBOCA SEM-DISNET revolves around the concept of disease, modeled in `ncit:C7057` class, reused from the NCIt. The semantic type is specified by the class `sio:SIO_000326`, while the disease class (meaning categorizations of diseases) is defined under `obo:HP_0000118`. Disease markers are modeled with `ncit:C18329`. The `ncit:C7057` class is associated to classes that represent relationships between concepts, including the associations with non-coding RNAs (`ncit:C26549`) and with Orphanet classification `ordo:Orphanet_557492` associations. Other classes modeling the relationships with diseases include:

- **Disease-gene association.** Modeled with `sio:SIO_000983`, it is quantified by the NCIt class `ncit:C25338` (labeled 'Score'). The `ncit:C16612` class (labeled 'Gene'), is related with WikiPathways metabolic `wp:Pathway`, the associated `obo:SO_0001060` ('Variant') and the encoded `ncit:C17021` ('Protein'). Each of these proteins belongs to an `obo:PR_000000001` (referring to a 'Protein class') and participates in `ncit:C18469` ('Protein-Protein Interactions' or PPIs). Both proteins and genes are related to the class `ncit:C14250` ('Organism') and proteins and organisms can act as a `cco:Target`, class reused from the ChEMBL ontology.
- **Disease-variant association.** It has an associated score defined by the mentioned 'Score' class. The `obo:SO_0001060` class (representing 'Variant') is reused from OBO Sequence Ontology (SO).
- **Disease-phenotype association.** This association and the related class `Phenotype` have been modeled by SEM-DISNET, since no classes were found in other ontologies with the exact meaning and specifications corresponding to the one in DISNET database.
- **Disease-drug association.** The association between diseases and drugs was modeled reusing a class from CTD, `ctd:Chemical-Disease-Association`. Drug-disease pairs were classified into three created subclasses depending on the effect that the drug has on the disease. It can act as a marker if the chemical correlates with a disease (`DrugDiseaseMarker`), it can be used to treat the disease (`DrugDiseaseTherapeutic`), or the association might have been inferred (`DrugDiseaseInferred`).

The `cco:Drug` class was reused from the ChEMBL ontology and represents any substance which when absorbed into a living organism may modify one or more of its functions. Besides its associated properties, it references to classes which represent relationships:

---

[14] https://medal.ctb.upm.es/internal/gitlab/disnet/sem-disnet/blob/master/diagrams/SEM-DISNET_hierarchy.png.

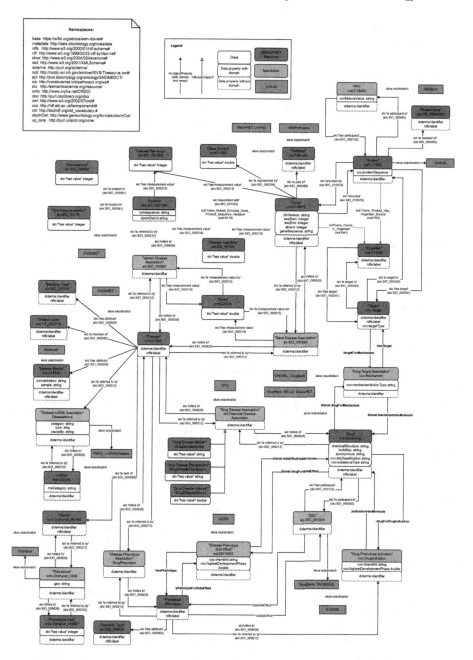

Fig. 2. Conceptualization of the EBOCA SEM-DISNET module.

– **Drug-target association.** Corresponding to `cco:Mechanism`, it is known as the mechanism of action of a drug when it addresses to `cco:Target`.

- **Drug-phenotype association.** The two entities `cco:Drug` and `Phenotype` are connected by three association classes referring to the possible causality of the interaction:
  - **Indication.** Reused from ChEMBL class `cco:DrugIndication`, it is referred to the approved uses of a medication.
  - **Side effect.** Association between symptoms considered as adverse effects that cause patient phenotype changes in response to the treatment with a drug. It was taken from SNOMED CT database (`sct:662014003`).
  - **Drug-drug interaction (DDI).** The class `sio:SIO_001006` was reused to model these interactions. DrugBank is the main source of data providing DDIs in DISNET, but their redistribution is not allowed. However, DISNET also holds TWOSIDES information, which also provides the potential adverse side effects that the DDI might cause.

EBOCA SEM-DISNET ontological model is represented in Fig. 2. The resource classes, represented by blue rounded boxes, correspond to biomedical entities; while association classes, which establish the relationship between two biomedical entities, are represented in orange. Attributes correspond to the white rounded boxes, and linkouts to external original sources to green boxes.

4.2   EBOCA Evidences

The EBOCA Evidences module[15] is built to enrich the EBOCA SEM-DISNET module. It is focused on providing metadata and provenance information about the associations between biomedical concepts. These evidences of associations may come from well known curated sources, or may be extracted or inferred from texts. This module enables describing in more detail the type of evidence supporting the association, the agents involved in its extraction and publication, and if applicable, the texts and sources where the evidence is extracted from.

This module reuses the following ontologies: SIO [13] for associations; Evidence & Conclusion Ontology (ECO) [17] for the different kinds of evidences; Friend Of A Friend (FOAF) [19] for creators of evidences; DCMI Metadata Terms[16] for general metadata; Provenance, Authoring and Versioning (PAV) [8] for metadata of evidences; and the SPAR Ontologies [28] FRBR-aligned Bibliographic Ontology (FaBiO), Document Components Ontology (DoCO) to describe and expression of work and its parts, paragraphs.

The conceptualization of the EBOCA Evidences module is depicted in Fig. 3. The link to the SEM-DISNET module is the class `sio:SIO_000897` ("Association"), which is the superclass of all kinds of associations represented in the other module. These associations may have a supporting `Evidence`, which in turn can be inferred (`ECO:0007672` "computational inference") or extracted from curated sources (`ECO:0006151` "documented statement evidence"). The class `Evidence` also enables describing the date of creation and update, version, software and agent responsible for its creation, and its provenance.

---

[15] https://w3id.org/eboca/evidences.

[16] https://www.dublincore.org/specifications/dublin-core/dcmi-terms/.

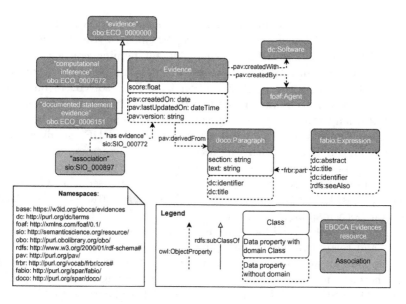

Fig. 3. Conceptualization of the EBOCA Evidences module.

This module is focused on describing paragraphs from curated resources and natural text, as it is the main use case. Thus, this module represents how evidences are derived from (`pav:derivedFrom`) `doco:Paragraph`, which is part of `fabio:Expression`, a class that describes, among others, papers. Both paragraphs and expressions can be uniquely identified, and include more attributes to describe them (e.g. the section paragraphs belong to and complete text; title, abstract and URL for papers). Evidences can also be derived from other sources, case in which the resource URI is specified with `pav:derivedFrom` too.

The scope of this module is limited for now to describe provenance of evidences when extracted from texts and curated resources. As the EBOCA resource grows and adds more different kinds of evidence extractions methods, it will evolve to represent them. The specific requirements for EBOCA Evidences are publicly available[17]. These requirements are tested in Themis and with SPARQL queries, as Sect. 5 explains in detail.

5   Ontology Evaluation

Both EBOCA modules are evaluated in different ways to identify inconsistencies, pitfalls, errors and to check that they meet the requirements.

The modelling pitfalls were evaluated with the OntOlogy Pitfall Scanner! (OOPS!) [30]. As many ontologies are reused in both modules, some pitfalls appear, but all of them pointing to reused ontology entities. The pitfall record

---

[17] https://drugs4covid.github.io/EBOCA-portal/requirements/requirements-evidences.html.

includes P04 (Creating unconnected ontology elements), P08 (Missing annotations), P11 (Missing domain or range in properties), P13 (Inverse relationships not explicitly declared) and P22 (Using different naming conventions in the ontology). However, all pitfalls that appeared in the modules for newly implemented entities during the implementation have been solved. Inconsistencies were checked with the reasoners Pellet and HermiT. For both modules, no errors were found when running the mentioned reasoners. To ensure that the modules met the requirements, test data was transformed into RDF, building a small Knowledge Graph (KG), which was used to run the corresponding SPARQL queries. This transformation was performed separately in each module.

To build each KG, declarative mapping languages are used: the W3C Recommendation R2RML [10] for EBOCA SEM-DISNET, and RML [12] for EBOCA Evidences. Both mappings have been created using Mapeathor [20]. The mapping engine Morph-KGC [1] has been used to materialize both KGs. The resulting KGs and related resources (mappings, queries) are available in GitHub[18]. The mappings can also be used for virtualization, but for evaluating the ontology we chose the materialization approach (i.e. generating the RDF triples).

For EBOCA SEM-DISNET, as the original relational database scales to large amounts of data, we decided to use a subset of the entirety of DISNET information to generate the KG. We included several concepts and some of their associations. Those concepts were diseases, genes, drugs and pathways. As for the associations, the data corresponding to disease-gene, disease-variant, disease-drug, gene-variant, gene-pathway and drug-drug associations were transformed to RDF. Despite the fact that it was just a fragment, this part of the KG contained 8,691,974 triplets, weighing a total of 1.6 GB. The graph was uploaded to Blazegraph[19], which enabled the successful execution of 15 SPARQL queries. They represent the different competency questions, validating the variety of concepts and associations in EBOCA SEM-DISNET module. The queries were targeted to both the modeled ontology as well as to the RDF instances (named individuals) created and included in the KG.

For EBOCA Evidences, JSON data coming from biomedical concepts extracted from 5,000 paragraphs of the CORD-19 corpus were used to build the test KG. A Named Entity Recognition and Normalization system (BioNER+ BioNEN), based on fine-tuned BioBERT models [2], was used to identify diseases, drugs and genetic terms. Thus, associations from the extracted terms were transformed and annotated with metadata about their evidence and provenance. The queries were run in Blazegraph with success. The exception was query `eboca-ev7`, that cannot output any results because of the extraction method used outputs no confidence score. However, to check the ontology consistency independently of the test data, the requirements formulated as statements were also validated using Themis [15]. Themis is a web service that enables executing tests to ontologies. These tests correspond to the ontology requirements, have been tested with success.

---

[18] https://github.com/drugs4covid/EBOCA-Resources.
[19] https://blazegraph.com/.

Evaluations for both modules were run successfully, with good results from OOPS!, SPARQL queries and Themis. This ensures that the ontology is consistent, meets the requirements, and to the best of our knowledge, has no errors.

6 Conclusions

This work presents EBOCA, an ontology that models evidences and provenance of associations between biomedical concepts. It is composed of two modules, (i) EBOCA SEM-DISNET for biomedical concepts and associations, and (ii) EBOCA Evidences for evidence, provenance and metadata information of associations. EBOCA aims to put forward a resource to model biomedical concepts and their association with the possibility of tracing them via evidences. We have explained LOT, the methodology employed for the development of these two modules; we have described them, and how their evaluation was performed to validate them. We conclude that the proposed ontology is fit to its purpose, to represent biomedical entities, their associations and evidence supporting them.

As future work, we want to create a complete Knowledge Graph to integrate data from DISNET enriched with its evidence information, as well as new associations from unstructured text analysis. Moreover, the evidences will be published as Nanopublications to promote its reuse. Other applications we foresee from the use of EBOCA are the possibility of proposing drug candidates for new diseases by approaching the link prediction challenge in the graph, addressing a better understanding of diseases by means of knowledge reasoning, or detecting and managing contradictory evidences.

Acknowledgments. This work is supported by the DRUGS4COVID++ project, funded by Ayudas Fundación BBVA a equipos de investigación científica SARS-CoV-2 y COVID-19. The work is also supported by "Data-driven drug repositioning applying graph neural networks (3DR-GNN)" under grant "PID2021-122659OB-I00" from the Spanish Ministerio de Ciencia, Innovación y Universidades. LPS's work is supported by "Programa de fomento de la investigación y la innovación (Doctorados Industriales)" from Comunidad de Madrid (grant "IND2019/TIC-17159").

References

1. Arenas-Guerrero, J., Chaves-Fraga, D., Toledo, J., Pérez, M.S., Corcho, O.: Morph-KGC: scalable knowledge graph materialization with mapping partitions. Semantic Web 1–20 (2022). http://www.semantic-web-journal.net/system/files/swj3135.pdf
2. Badenes-Olmedo, C., Alonso, A., Corcho, O.: An overview of drugs, diseases, genes and proteins in the cord-19 corpus. Procesamiento del Lenguaje Natural, vol. 69 (2022)
3. Belleau, F., Nolin, M.A., Tourigny, N., Rigault, P., Morissette, J.: Bio2RDF: towards a mashup to build bioinformatics knowledge systems. J. Biomed. Inform. **41**(5), 706–716 (2008). https://doi.org/10.1016/j.jbi.2008.03.004
4. Bodenreider, O.: The Unified Medical Language System (UMLS): integrating biomedical terminology. Nucleic Acids Res. **32**(suppl_1), D267–D270 (2004). https://doi.org/10.1093/nar/gkh061

5. Bodenreider, O., Mitchell, J.A., McCray, A.T.: Biomedical ontologies. In: Pacific Symposium on Biocomputing, pp. 76–78 (2005)

6. Bodenreider, O., Stevens, R.: Bio-ontologies: current trends and future directions. Brief. Bioinform. **7**(3), 256–274 (2016). https://doi.org/10.1093/bib/bbl027

7. Chávez-Feria, S., García-Castro, R., Poveda-Villalón, M.: Chowlk: from UML-based ontology conceptualizations to owl. In: Groth, P., et al. (eds.) The Semantic Web, pp. 338–352. Springer, Cham (2022). https://doi.org/10.1007/978-3-031-06981-9_20

8. Ciccarese, P., Soiland-Reyes, S., Belhajjame, K., Gray, A.J., Goble, C., Clark, T.: Pav ontology: provenance, authoring and versioning. J. Biomed. Semant. **4**(1), 1–22 (2013)

9. Consortium, G.O.: The gene ontology (GO) database and informatics resource. Nucleic Acids Res. **32**(suppl_1), D258–D261 (2004)

10. Das, S., Sundara, S., Cyganiak, R.: R2RML: RDB to RDF Mapping Language, W3C Recommendation, 27 September 2012. www.w3.org/TR/r2rml

11. Davis, A.P., et al.: Comparative toxicogenomics database (CTD): update 2021. Nucleic Acids Res. **49**, D1138–D1143 (2021). https://doi.org/10.1093/nar/gkaa891

12. Dimou, A., Sande, M.V., Colpaert, P., Verborgh, R., Mannens, E., Van De Walle, R.: RML: a generic language for integrated RDF mappings of heterogeneous data. In: LDOW (2014)

13. Dumontier, M., et al.: The semanticscience integrated ontology (SIO) for biomedical research and knowledge discovery. J. Biomed. Semant. **5**(1), 14 (2014). https://doi.org/10.1186/2041-1480-5-14

14. Eilbeck, K., et al.: The Sequence Ontology: a tool for the unification of genome annotations. Genome Biol. **6**(5), R44 (2005). https://doi.org/10.1186/gb-2005-6-5-r44

15. Fernández-Izquierdo, A., Cimmino, A., García-Castro, R.: Supporting demand-response strategies with the delta ontology. In: 2021 IEEE/ACS 18th International Conference on Computer Systems and Applications (AICCSA), pp. 1–8. IEEE (2021)

16. Garijo, D.: WIDOCO: a wizard for documenting ontologies. In: d'Amato, C., et al. (eds.) ISWC 2017. LNCS, vol. 10588, pp. 94–102. Springer, Cham (2017). https://doi.org/10.1007/978-3-319-68204-4_9

17. Giglio, M., et al.: Eco, the evidence & conclusion ontology: community standard for evidence information. Nucleic Acids Res. **47**(D1), D1186–D1194 (2019)

18. Goh, K.I., Cusick, M.E., Valle, D., Childs, B., Vidal, M., Barabási, A.L.: The human disease network. Proc. Natl. Acad. Sci. **104**(21), 8685–8690 (2007). https://doi.org/10.1073/pnas.0701361104

19. Graves, M., Constabaris, A., Brickley, D.: FOAF: connecting people on the semantic web. Cataloging Classif. Q. **43**(3–4), 191–202 (2007)

20. Iglesias-Molina, A., Pozo-Gilo, L., Doña, D., Ruckhaus, E., Chaves-Fraga, D., Corcho, Ó.: Mapeathor: simplifying the specification of declarative rules for knowledge graph construction. In: ISWC (Demos/Industry) (2020)

21. Jackson, R., et al.: OBO foundry in 2021: operationalizing open data principles to evaluate ontologies. Database **2021**, baab069 (2021). https://doi.org/10.1093/database/baab069

22. Köhler, S., et al.: The human phenotype ontology in 2021. Nucleic Acids Res. **49**(D1), D1207–D1217 (2021). https://doi.org/10.1093/nar/gkaa1043

23. Lagunes-García, G., Rodríguez-González, A., Prieto-Santamaría, L., del Valle, E.P.G., Zanin, M., Menasalvas-Ruiz, E.: DISNET: a framework for extracting phenotypic disease information from public sources. PeerJ **8**, e8580 (2020). https://doi.org/10.7717/peerj.8580
24. Lebo, T., et al.: PROV-O: The PROV ontology (2013). www.w3.org/TR/prov-o/
25. Martens, M., et al.: WikiPathways: connecting communities. Nucleic Acids Res. **49**(D1), D613–D621 (2021). https://doi.org/10.1093/nar/gkaa1024
26. Mendez, D., et al.: ChEMBL: towards direct deposition of bioassay data. Nucleic Acids Res. **47**, D930–D940 (2019). https://doi.org/10.1093/nar/gky1075
27. Natale, D.A., et al.: The Protein Ontology: a structured representation of protein forms and complexes. Nucleic Acids Res. **39**(suppl_1), D539–D545 (2011). https://doi.org/10.1093/nar/gkq907
28. Peroni, S., Shotton, D.: The SPAR ontologies. In: Vrandečić, D., et al. (eds.) ISWC 2018. LNCS, vol. 11137, pp. 119–136. Springer, Cham (2018). https://doi.org/10.1007/978-3-030-00668-6_8
29. Piñero, J., et al.: The DisGeNET knowledge platform for disease genomics: 2019 update. Nucleic Acids Res. **48**, D845–D855 (2020). https://doi.org/10.1093/nar/gkz1021
30. Poveda-Villalón, M., Gómez-Pérez, A., Suárez-Figueroa, M.C.: Oops!(ontology pitfall scanner!): an on-line tool for ontology evaluation. Int. J. Semant. Web Inf. Syst. (IJSWIS) **10**(2), 7–34 (2014)
31. Poveda-Villalón, M., Fernández-Izquierdo, A., Fernández-López, M., García-Castro, R.: LOT: an industrial oriented ontology engineering framework. Eng. Appl. Artif. Intell. **111**, 104755 (2022). https://doi.org/10.1016/j.engappai.2022.104755
32. Prieto Santamaría, L., Díaz Uzquiano, M., Ugarte Carro, E., Ortiz-Roldán, N., Pérez Gallardo, Y., Rodríguez-González, A.: Integrating heterogeneous data to facilitate COVID-19 drug repurposing. Drug Discovery Today **27**(2), 558–566 (2022). https://doi.org/10.1016/j.drudis.2021.10.002
33. Prieto Santamaría, L., Ugarte Carro, E., Díaz Uzquiano, M., Menasalvas Ruiz, E., Pérez Gallardo, Y., Rodríguez-González, A.: A data-driven methodology towards evaluating the potential of drug repurposing hypotheses. Comput. Struct. Biotechnol. J. **19**, 4559–4573 (2021). https://doi.org/10.1016/j.csbj.2021.08.003
34. Prieto Santamaría, L., García del Valle, E.P., Zanin, M., Hernández Chan, G.S., Pérez Gallardo, Y., Rodríguez-González, A.: Classifying diseases by using biological features to identify potential nosological models. Sci. Rep. **11**(1), 21096 (2021). https://doi.org/10.1038/s41598-021-00554-6
35. Queralt-Rosinach, N., Piñero, J., Bravo, A., Sanz, F., Furlong, L.I.: DisGeNET-RDF: harnessing the innovative power of the semantic web to explore the genetic basis of diseases. Bioinformatics **32**(14), 2236–2238 (2016)
36. Redaschi, N., Consortium, U.: UniProt in RDF: tackling data integration and distributed annotation with the semantic web. Nat. Precedings (2019). https://doi.org/10.1038/npre.2009.3193.1
37. Schriml, L.M., et al.: The human disease ontology 2022 update. Nucleic Acids Res. **50**, D1255–D1261 (2022). https://doi.org/10.1093/nar/gkab1063
38. Suárez-Figueroa, M.C., Gómez-Pérez, A., Fernandez-Lopez, M.: The neon methodology framework: a scenario-based methodology for ontology development. Appl. Ontol. **10**(2), 107–145 (2015)

39. García del Valle, E.P., Lagunes García, G., Prieto Santamaría, L., Zanin, M., Menasalvas Ruiz, E., Rodríguez-González, A.: DisMaNET: a network-based tool to cross map disease vocabularies. Comput. Methods Programs Biomed. **207**, 106233 (2021). https://doi.org/10.1016/j.cmpb.2021.106233

40. Vasant, D., et al.: ORDO: an ontology connecting rare disease, epidemiology and genetic data. In: Bio-Ontologies ISMB 2014, July 2014

41. Wishart, D.S., et al.: DrugBank 5.0: a major update to the DrugBank database for 2018. Nucleic Acids Res. **46**(D1), D1074–D1082 (2018). https://doi.org/10.1093/nar/gkx1037

42. Zahn-Zabal, M., et al.: The neXtProt knowledgebase in 2020: data, tools and usability improvements. Nucleic Acids Res. **48**, D328–D334 (2020). https://doi.org/10.1093/nar/gkz995

Counter Effect Rules Mining
in Knowledge Graphs

Lucas Simonne[1]([✉]), Nathalie Pernelle[2], and Fatiha Saïs[1]

[1] LISN, CNRS UMR 9015, Paris Saclay University, Orsay, France
{lucas.simonne,fatiha.sais}@lisn.fr
[2] LIPN, CNRS UMR 7030, Sorbonne Paris Nord University, Paris, France
pernelle@lipn.univ-paris13.fr

Abstract. Discovering causal relationships is the goal of many experiments in science. Such a relationship indicates that a variation in an attribute, *i.e.*, the *treatment*, implies a variation, *i.e.*, has an *effect*, on another attribute, *i.e.*, the *outcome*. Mining causal relationships have been studied in a recent approach in Knowledge Graphs, where *differential causal rules* are mined. Such rules express an effect of a treatment on a subset of instances described by a graph pattern named *strata*. However, these rules can be difficult to interpret, especially when a treatment has different effects depending on the strata it is expressed on. This paper presents *counter effect rules* that can be discovered from differential causal rules to facilitate their interpretation. This representation allows to point out the strata that lead to opposite effects for the same treatment. Our experiment shows that counter effect rules can be discovered on a real dataset.

Keywords: Knowledge Discovery · Causal Rules · Knowledge Graphs · Explainability

1 Introduction

Causal relationships are of great interest in a large diversity of domains, such as drug discovery in order to determine whether a drug has an expected effect or not. Although it is widely studied in tabular data [2], only a few approaches focus on causality in knowledge graphs (KGs). Recent papers were proposed [3,8] and are based on graphical models or on the potential outcome framework [7].

In this work, we focus on differential causal rules (DCR) that allow expressing that for two different class instances a difference in property values representing a treatment may explain a difference in a numerical property value representing a studied outcome [8]. For instance, a DCR could be that for two persons, a higher-calorie diet may explain a higher weight gain. This simple DCR can be valid for all instances of persons, but other DCRs can be valid for a subset

---

Supported by the DATAIA institute.

of instances only, *i.e.* referred as strata, that are represented by a basic graph pattern. The strata of a DCR makes the rule more expressive and allows showing that a causal relationship can be only valid on a specified sub-population.

In *Simonne et al.* [8], an experiment on a KG about people's willingness to reduce their meat consumption showed that the mined DCRs were judged meaningful by domain experts. However, it has been shown that the discovery of DCRs with strata can lead to very specific rules that are not ordered and are sometimes difficult to interpret. Experiments have shown that sometimes a treatment can be observed to have an opposite effect depending on the strata, this knowledge is of great interest to experts. For instance, some rules could express that living in the countryside compared to living in a city increases people's willingness to reduce their meat consumption when people are women, but decreases it when people are men.

This paper presents a new method that allows mining a new type of rules that we name Counter Effects Rules (CER). CERs are obtained by processing DCRs where treatments have opposite effects on different strata. Not only CERs are more interpretable, but they can bring new knowledge to experts by ranking treatments and displaying the local effects of treatments.

2   Related Work

Causality discovery and rules mining are both extensively studied in different fields, such as machine learning and statistics [2] for causality discovery, and relational databases for rules mining [1,4]. As we deal with KGs with heterogeneous and incomplete information, such works are not relevant to our problem. We, therefore, restrict the related works to ones that consider KGs.

Differential Causal Rules. In a recent approach, *Simonne et al.* [8] uses a widely used causal discovery framework, the potential outcome framework, to determine DCRs. This framework [7] states that the effect of a treatment can be obtained by comparing the outcome of an instance to its counterfactual. The counterfactual is the outcome of an instance would it have had another treatment value. The comparison of pairs composed of an instance and its counterfactual is then used to obtain the effect of a treatment, and to mine DCRs [8]. As mentioned in Sect. 1, while DCRs are expressive, their interpretation becomes difficult when a treatment has different effects on different strata.

Association Rules. Association rules mining approaches are commonly used to discover knowledge or to remove erroneous triples [1,4]. Although an association rule represents an association, it does not necessarily indicate a causal relationship. Causal relations are obtained by considering the similarity of instances while this is not done in association rule mining approaches. A naive approach could consist of setting the head of an association rule to a difference of values on a path between two instances as a parameter in the rule mining algorithm. However, the bodies of the resulting rules would not be suited for explaining this difference. For instance, the bodies could express partial descriptions of the instances, and/or different attributes [8].

Counterfactuals Explanations and Action Rules. Counterfactuals explanations (CE) explain how to obtain an alternative decision from a machine learning model by identifying changes in input features that lead to a desired outcome from the model [6]. Counterfactual explanation approaches can be used for interpretability purposes to understand how a model makes predictions or for a user to know what input to use to obtain a certain outcome. Action Rules [5,9] are a part of counterfactuals explanations and are an extension of classification rules. They express that an action, *i.e.* a change in a variable, will change the class of an instance described by a rule. While such rules and counterfactuals explanations in general focus on explaining different outcomes of machine learning models, they do not study why the same treatment has opposite effects for instances within a dataset. Moreover, as action rules are generally discovered from association rules previously mined, they do not consider instances' similarity and therefore do not indicate a causal effect.

3   Preliminaries and Problem Statement

In this section, we first give our definition of a knowledge graph, then remind a set of important definitions that have been introduced in [8], and finally give the definition of *Counter Effect Rules* on which the present work is focusing on.

Definition 1 (Knowledge Graph). *We consider a knowledge graph KG defined by a pair $(\mathcal{O}, \mathcal{F})$ where $\mathcal{O}$ is an ontology represented in OWL and $\mathcal{F}$ is a set of RDF triples describing class instances of $\mathcal{O}$.*

Differential Causal Rule (DCR) [8]. A DCR is defined as a rule expressed in first order logic in the form of: $strata \wedge treatment \Rightarrow outcome$. Each rule is applied on two instances of a *strata*, where a *treatment* explains an *effect*.

Strata [8]. A strata ST is a basic graph pattern in RDF that represent all instances that have its pattern in their descriptions, such that a DCR with a strata ST is applied on instances from ST. An example strata is $ST_1(X)$: $isA(X, man) \wedge livesIn(X, city)$, that gathers men living in cities.

A treatment T [8]. A treatment indicates that two instances have different values on a numerical or a categorical path, referred to as the property path.

An outcome O [8]. An outcome indicates that two instances have different values on a numerical path, referred to as the outcome path.

Example 1. Given two instances X_1 and X_2, a treatment T_1 indicates that X_1 has an omnivorous diet and X_2 a vegetarian diet, and an outcome O_1 indicates that X_1 has a higher will to reduce its meat consumption than X_2. The rule DCR_1 can be build using the previous examples and express that, for men living in cities, being omnivorous compared to being vegetarian explains a higher will to reduce its meat consumption.

Below we give our definition of Counter Effect Rules (CER) using the terms previously introduced.

Definition 2 (Counter Effect Rules). *A counter effect rule CER_T expresses that, for two instances X_1 and X_2 of a class $c \in \mathcal{O}$, a treatment T explains an outcome O if X_1 and X_2 belong to a strata ST_i, while T explains the opposite outcome $\overline{O}$ if X_1 and X_2 belong to a strata $ST_j \neq ST_i$. More precisely, a change in the strata from ST_i to ST_j, that we denote by an arrow $\rightarrow$, explains an opposite effect of a treatment. More formally, a CER_T is defined as follows:*

$$CER_T(X_1, X_2) : c(X_1) \wedge c(X_2) \wedge T(X_1, X_2) \wedge P_O(X_1, U_1) \wedge P_O(X_2, U_2) \wedge$$
$$[ST_i(X_1) \wedge ST_i(X_2) \rightarrow ST_j(X_1) \wedge ST_j(X_2)] \Rightarrow [O(X_1, X_2) \rightarrow \overline{O}(X_1, X_2)] \quad (1)$$

Example 2. Let $ST_2 : isA(X, man) \wedge livesIn(X, campaign)$, DCR_2 express that treatment T_1 has the opposite effect of O_1 on the instances with ST_2, *i.e.*, omnivorous people have a lower willingness to reduce their meat consumption compared to vegetarian people. The resulting CER is:

$$hasDiet(X_1, omnivorous) \wedge hasDiet(X_2, vegetarian) \wedge reduceMeat(X_1, u_1) \wedge$$
$$reduceMeat(X_2, u_2) \wedge [ST_1(X_1) \wedge ST_1(X_2) \rightarrow ST_2(X_1) \wedge ST_2(X_2)]$$
$$\Rightarrow [higherThan(u_1, u_2) \rightarrow lessThan(u_1, u_2)].$$

4   Discovering Counter Effects Rules

The approach we developed for discovering CERs is performed in two steps. In the first step, differential causal rules are discovered by using the approach developed by *Simonne et al.* [8]. In the second step, the differential causal rules are processed to obtain counter effects rules.

4.1   Step 1: Mining Differential Causal Rules

DCRs are mined by applying the method from *Simonne et al.* [8]. It considers a KG, a target class c, the set of property paths $\mathcal{P}$ that lead to possible treatments P_T, the property path leading to the outcome P_O and expert knowledge to guide the mining.

This algorithm discovers DCRs using a bottom-up approach. In a first step, it generates potential DCRs and evaluates them using a metric inspired by the oddsratio. In [8], the obtained DCRs are said to be *specific* as they express treatments on stratas composed of several predicates. In a second step, the strata of the DCR are merged to obtain more general rules if they share the same treatment and outcome. While the effect of a specific rule is quantified by the oddsratio, a rule obtained from a merge is not quantified. As for this new approach, only the specific DCR are considered as input to the next step.

4.2   Step 2: Obtaining Opposite Effects

In the second step of the algorithm, the set of CERs, $\mathcal{D}$, is mined from the DCRs previously mined. A CER is obtained from two specific DCR if they have the same treatment but opposite outcomes.

As shown in Eq. 1, a CER expresses that a change from one strata to another can explain an opposite effect. Instead of outputting all the CERs obtained by crossing the DCRs, we introduce a threshold d_c that is the maximum number of changes a rule can have, *i.e.*, the number of differences between the two strata it is built from. By doing so, we ensure the generation of the best set of rules that can be interpreted. More precisely, the DCRs are first sorted on the treatment they express. Then, for each treatment T, two sets of rules are obtained: $\mathcal{R}_{(T,O)}$ and $\mathcal{R}_{(T,\overline{O})}$ depending on the rules outcome. A potential counter effect rule per pair $(R_i, R_j) \in \mathcal{R}_{(T,O)} \times \mathcal{R}_{(T,\overline{O})}$ is constructed, and, with ST_i the strata of R_i, is added to $\mathcal{D}$ if its number of changes $n(ST_i - ST_l) \leq d_c$. Finally, the CER from $\mathcal{D}$ are merged to obtain more general rules.

Example 3. Let us consider DCR_1 and DCR_2 the DCR given in Sect. 3. DCR_1 and DCR_2 have the same treatment but opposite outcomes and can therefore be used to generate a potential CER, CER_1, described in Example 2. With d_c set to 1, as $n(ST_1 - ST_2) = 1$, CER_1 can be added to $\mathcal{D}$.

5   Experiments

We conducted experiments in order to show the relevance of the CERs that can be mined over a knowledge graph. We compared them to both the DCRs mined using the approach of *Simonne et al.* [8] and to association rules mined using *AMIE* [1].

Dataset. The dataset *Vitamin* has been used for the experiment. It describes people and their socioeconomic characteristics such as their age, gender, current and ideal diet, opinions on animal welfare facts and climate change. It has more than 86k triples and 1714 instances of c the target class *Person*.

The outcome is the difference between a person's current and ideal diet, *i.e.*, their willingness to reduce their meat consumption. It is denoted by $P_O = reduceMeat$ having values $\in \mathbb{N}$. We are interested in observing if treatments can have opposite effects and why, *i.e.*, if some instances have higher *reduceMeat* values, while other have lower *reduceMeat* values for a same treatment.

Results. The set of CERs has been obtained by using the DCRs mined by the approach of [8] on *Vitamin* by using the same parameters as mentioned in [8]. d_c is set to 1. Let $\mathcal{P} = \{opinion, gender, livesIn, age, education\}$.

1) Mined CERs. In total, 23 CERs with a change of one element were discovered using the set of 141 specific DCRs previously mined. We remind the reader that the DCRs used to create the CER were judged accurate and meaningful by experts of the field. An example of a CER, CER_2, is presented in Eq. 2[1], where men that live in cities have a higher will to reduce their meat consumption than men living in campaigns, had they attended university, while this will is higher

---

[1] The value 1 for predicate *hasOpinion* represents a person with pro-meat opinions who believes that breeding is unrelated to climate issues.

for men living in campaigns compared to men living in cities for men who did not attend university. Table 1 presents an extract of treatments involved in the mined DCRs (*Treatment*), the number of DCRs they are involved in (*#DCR*) and the percentage of DCRs having the same outcome (*% same O*), the number of CERs (*#CER*), and the most common treatments that explain their counter effect (*Counter Treatment*).

$$CER_2 : [gender(X, man) \wedge hasOpinion(X, 1)] \wedge livesIn(X_1, city) \wedge$$
$$livesIn(X_2, campaign) \wedge reduceMeat(X_1, u_1) \wedge reduceMeat(X_2, u_2) \wedge$$
$$[\textbf{education}(\textbf{X}, \textbf{university}) \rightarrow \textbf{education}(\textbf{X}, \textbf{high\_school})] \qquad (2)$$
$$\Rightarrow [\textbf{greaterThan}(\textbf{u}_1, \textbf{u}_2) \rightarrow \textbf{greaterThan}(\textbf{u}_2, \textbf{u}_1)]$$

Table 1. Extract of results. X_2 has a higher will to reduce its meat consumption than X_1.

Treatment	#DCR (% same O)	#CER	Counter treatment
$gender(X_1, man) \wedge gender(X_2, woman)$	10 (100%)	0	*NA*
$livesIn(X_1, city) \wedge livesIn(X_2, campaign)$	18 (88%)	2	gender, education
$education(X_1, university) \wedge education(X_2, high\_school)$	16 (50%)	7	gender, opinion

2) Comparison to DCRs. Treatments can explain the same outcome in all DCRs they are expressed in, as for the treatment on gender (Line 1, 100%), or opposite outcomes, as for the treatment on education (Line 3, 50%).

Specific DCRs where treatments have the same outcome can be generalised and are easily interpreted. They show relations that are already known and published in the literature [10]. However, DCRs with treatment having opposite outcomes are difficult to interpret. The set of CERs is of interest for these rules. First, it allows their analysis by comparing relevant DCRs and showing that a treatment can have different effects and why. Secondly, in opposition to treatments with a consistent effect on all strata, the CERs display treatments that have local effects. They can therefore be used to discover knowledge that was less likely to be discovered with DCRs, and to guide a deeper analysis of such local effects, *e.g.*, a missing variable or an unknown event.

In addition to analysing the effect of a given treatment, the set of CERs can be used to rank treatments. The ranking is based on (i) how consistent a treatment is, *i.e.*, if its effect is the same in all rules it is expressed in, and (ii) how often a treatment changes the effect of another. For instance, the treatment on path *gender* seems the stronger in Table 1 as its effect is consistent in every DCR it appears in, and is responsible for a change of effect of other treatments. On the opposite, the treatment on path *education* seems to be the weaker as it explains an outcome in half of its DCR, and the opposite outcome in the other half. The study of both sets of rules is therefore complementary.

3) Comparison to Association Rules. A set of association rules has been mined using AMIE [1] with a support of 100 and confidence of 0.1. The target head of the rules has been set by introducing a new predicate that compares the difference in the *reduceMeat* predicates. Action rules were built by using the association rules. A set of 111 rules were mined using AMIE. However, none of them was able to explain a difference in the *reduceMeat* predicate between two instances. More precisely, while the body of the mined rules could contain predicates on both instances, they do not compare the same ones for the two instances. In consequence, a naive approach that compares association rules with heads having opposite predicates would not discover meaningful rules.

6   Conclusion

In this paper, we proposed an approach that mines counter effect rules. These rules express treatments that can have opposite effects within instances of a KG. A preliminary experiment shows the interest in our approach, especially to rank treatments. Such rules are to be used in complement to causal rules discovery.

References

1. Galárraga, L.A., Teflioudi, C., Hose, K., Suchanek, F.M.: AMIE: association rule mining under incomplete evidence in ontological knowledge bases. In: 22nd International World Wide Web Conference, WWW 2013, Rio de Janeiro, Brazil, 13–17 May 2013, pp. 413–422 (2013)
2. Guo, R., Cheng, L., Li, J., Hahn, P.R., Liu, H.: A survey of learning causality with data: problems and methods. ACM Comput. Surv. **53**(4), 1–36 (2020)
3. Munch, M., Dibie, J., Wuillemin, P., Manfredotti, C.E.: Towards interactive causal relation discovery driven by an ontology. In: International Florida Artificial Intelligence Research Society Conference (2019)
4. Ortona, S., Meduri, V.V., Papotti, P.: RuDiK: rule discovery in knowledge bases. Proc. VLDB Endow. **11**, 1946–1949 (2018)
5. Ras, Z.W., Wieczorkowska, A.: Action-rules: how to increase profit of a company. In: Proceedings of the 4th European Conference on Principles of Data Mining and Knowledge Discovery, PKDD 2000, pp. 587–592 (2000)
6. Rasouli, P., Yu, I.C.: Care: coherent actionable recourse based on sound counterfactual explanations (2021). https://arxiv.org/abs/2108.08197
7. Rubin, D.: Estimating causal effects of treatment in randomized and nonrandomized studies. J. Educ. Psychol. B **66**(5), 688–701 (1974)
8. Simonne, L., Pernelle, N., Saïs, F., Thomopoulos, R.: Differential causal rules mining in knowledge graphs. In: Proceedings of the 11th on Knowledge Capture Conference, K-CAP 2021, New York, NY, USA, pp. 105–112 (2021)
9. Sýkora, L., Kliegr, T.: Action rules: counterfactual explanations in Python. In: CEUR Workshop Proceedings, vol. 2644, pp. 28–41 (2020)
10. Thomopoulos, R., Salliou, N., Taillandier, P., Tonda, A.: Consumers' motivations towards environment-friendly dietary changes: an assessment of trends related to the consumption of animal products. In: Leal Filho, W., Djekic, I., Smetana, S., Kovaleva, M. (eds.) Handbook of Climate Change Across the Food Supply Chain. Climate Change Management, pp. 305–319. Springer, Cham (2022). https://doi.org/10.1007/978-3-030-87934-1_17

A FAIR Core Semantic Metadata Model for FAIR Multidimensional Tabular Datasets

Cassia Trojahn[1]([✉])[iD], Mouna Kamel[1][iD], Amina Annane[2][iD],
Nathalie Aussenac-Gilles[1][iD], and Bao Long Nguyen[1]

[1] IRIT, Université de Toulouse, CNRS,
Université Toulouse 2 Jean Jaurès, Toulouse, France
`Cassia.Trojahn@irit.fr`
[2] Geotrend, Toulouse, France
`amina.annane@geotrend.fr`

Abstract. Tabular format is a common format in open data. However, the meaning of columns is not always explicit which makes if difficult for non-domain experts to reuse the data. While most efforts in making data FAIR are limited to semantic metadata describing the overall features of datasets, such a description is not enough to ensure data interoperability and reusability. This paper proposes to reduce this weakness thanks to a (FAIR) core semantic model that is able to represent different kinds of metadata, including the data schema and the internal structure of a dataset. This model can then be linked to domain-specific definitions to provide domain understanding to data consumers.

1 Introduction

Tabular format structures data into columns and rows. Each row provides values of properties of what is described by the row. Cells within the same column provide values for the same property. Columns can characterize dimensions within a multidimensional view. According to [2], the tabular format is the most widespread format for publishing data on the Web (37% of the datasets indexed by Google are in CSV or XLS). Data in this format has been made available as open data and datasets on various open data portals. On these portals, however, these datasets are described and presented with properties that are relevant to domain experts but not properly understood and reusable by other communities. For the latter, one of the challenges is to find relevant data among the increasingly large amount of continuously generated data, by moving from the point of view of data producers to the point of view of users and usages.

One way to overcome these weaknesses is to guarantee compliance of data to the FAIR principles: Findability, Accessibility, Interoperability, and Reusability [15]. These principles correspond to a set of 15 recommendations that aims to facilitate data reuse by humans and machines. They are domain-independent and

O. Corcho et al. (Eds.): EKAW 2022, LNAI 13514, pp. 174–181, 2022.
https://doi.org/10.1007/978-3-031-17105-5_13

may be implemented principally by: (F), assigning unique and persistent identifiers to datasets, and describing them with rich metadata that enable their indexing and discovery; (A), using open and standard protocols for dataset access; (I), using formal languages, and (FAIR) vocabularies to represent (meta)data; and (R), documenting (meta)data with rich metadata about usage license, provenance and data quality. So the first step towards the fulfilment of FAIR principles is to define precise metadata schemes. Indeed, 12 out of the 15 FAIR principles refer to metadata [15]. To go a step further in improving data FAIRness, several authors have shown that metadata schemes should be based on semantic models (i.e., ontologies) for a richer metadata representation [5]. Thanks to their ability to make data types explicit, in a format that can be processed by machines, ontologies are essential to make data FAIR [6].

While most efforts in data FAIRification are limited to specific kinds of metadata, mainly those describing the overall features of datasets, such a description is not enough to fully address all FAIR principles [7], in particular for promoting reuse of their data by other scientific communities. This paper addresses this challenge by proposing a core semantic model capable of representing different types of metadata, including the data schema and the internal structure of a dataset. This core model can be used in different domains and can be linked to domain-specific definitions to provide domain understanding for data consumers. The proposed model relies on existing FAIR vocabularies and ontologies and is itself compliant with the FAIR principles.

The rest of this paper is organised as follows. Section 2 discusses the main related work, followed by the description of the reused vocabularies and ontologies in Sect. 3. Section 4 presents the proposed model and Sect. 5 reports its evaluation. Finally Sect. 6 concludes the paper.

2   Related Work

A number of vocabularies has been proposed so far to represent metadata in general (Dublin core, VoID, Schema.org, DCAT, DCAT-AP), with extensions for accommodating specific kinds of data, such as geo-spatial data (GeoDCAT-AP) or statistical data (StatDCAT-AP). Several works also proposed specific metadata vocabularies. This is the case of [10] which presents a data model for generating ontology-based semantic metadata for spatial and temporal data, or of the European Open Science Cloud (EOSC) initiative[1] in the context of social sciences and humanities. Concerning observational data, several proposals have used RDF Data Cube (qb) combined to other vocabularies. In [9] the authors combined qb and SOSA to represent 100 years of temperature data in RDF. More recently, metereological data was represented in RDF with a semantic model that reused a network of existing ontologies (SOSA/SSN, Time, QUDT, GeoSPARQL, and qb) [16]. Instead, we propose here a core model that is common to different types of tabular data and that can be extended according to the specifics of the datasets (for instance, using QUDT, SOSA, GeoSPARQL).

---

[1] https://ddialliance.org/learn/what-is-ddi (accessed on 10th June 2022).

The work done here is an evolution of a previous one presented in [1]. While that first model had a special focus on representing spatio-temporal data (using GeoDCAT-AP and QB4st), here we propose a more generic core semantic model for representing any kind of tabular data for any domain by adopting DCAT and qb. Furthermore, we have introduced new notions such as the notion of a slice that can be also considered as a dataset, and the notion of collection of tabular data, as further detailed in Sect. 4.

3   Reusing Existing Vocabularies

The proposed model was developed following the NeOn methodology scenario 3 "*Reusing ontological resources*" [13]. We introduce here the main existing vocabularies that we relied on to build the core model, without detailing each activity of the methodology. These vocabularies provide metadata describing general features of datasets (DCAT), as well as the internal structure of a dataset (RDF data Cube and CSVW). All these vocabularies are recommended at least by the W3C or FairSharing[2] and thus act as FAIR vocabularies.

DCAT (Data CATalogue vocabulary). DCAT[3] is an RDF vocabulary designed to describe the datasets and data services in a catalog, thus facilitating the aggregation of metadata from multiple catalogs published on the web. It is based on 6 main classes (`Catalog`, `Resource`, `Dataset`, `Distribution`, `DataService` and `CatalogRecord`). It incorporates terms from existing vocabularies, including FOAF (relationships that people maintain with each other), PROV-O (provenance information), Dublin Core (metadata terms including properties, vocabulary encoding schemes, syntax encoding schemes and classes), SKOS (basic structure and content of concept schemes of controlled vocabulary) and vCard (people and organisations). DCAT was standardized in 2014 and has acquired the status of W3C recommendation.

RDF data Cube (qb). qb[4] is a W3C vocabulary dedicated to the representation of multidimensional data or hyper-cubes [14]. It builds upon several existing and recommended RDF vocabularies, such as SKOS, VoID (metadata about RDF datasets, intended to serve as a bridge between publishers and users of RDF data), Dublin Core, SCOVO (representation of statistical data), FOAF and ORG (organizational structures). qb allows the selection (i) of subsets of observations thanks to the notion of `Slice`, and (ii) of subsets of a given slice when the key slice has been fixed. Thus the publisher can identify and label those particular subsets. qb also allows the structure of a dataset to be described using the `DataStructureDefinition` and `ComponentProperty` entities. Each component property can be linked to the concept it represents (modelled as a SKOS concept).

---

[2] https://fairsharing.org/ (accessed on 10th June 2022).

[3] https://www.w3.org/TR/vocab-dcat-2/ (accessed on 8th June 2022).

[4] https://www.w3.org/TR/eo-qb/ (accessed on 8th June 2022).

CSVW. Resulting from the work of a W3C group on tabular data, CSVW[5] provides metadata at various levels, from table to groups of tables and how they are related to each other. A `Table` can be described with its url, schema, number of columns, foreign keys, transformations in other formats, etc.; each `Column` is then described by its name, title, type, position, whether the value is mandatory, etc. Furthermore, the interdependence between two tables may be represented by linking a column (or a set of columns) of a given table to a column (or a set of columns) of another table, thanks to references to their `ForeignKey`.

4   Proposed Model

We propose the Core Dataset Metadata Ontology (*dmo-core*) as a core semantic model capable of representing various types of metadata, including the data schema and the internal structure of tabular datasets (Fig. 1). It is a domain-independent model that can be enriched and specialized with domain ontologies to describe datasets in that domain. It is based on the FAIR vocabularies presented above. *dmo-core* is available online at https://w3id.org/dmo.

The notion of catalog is represented by `dcat:Catalog`, as a curated collection of metadata about resources (e.g., datasets and data services in the context of a data catalog). A catalog associates `dcat:Dataset`, which can be described with several types of metadata [4] using DCAT or DCT properties: descriptive metadata (`dct:description`, `dct:title`, `dcat:keywords`, etc.), quality (`dct:conformsTo`, etc.), provenance (`dct:publisher`, `dct:creator`), access rights (`dct:accessRights`, etc.) and versioning (`dct:hasVersion`, etc.). A `dcat:Dataset` may have different distributions `dcat:Distribution` (described by `dct:format`, `dcat:accessURL`, etc.), some of which may be in a tabular format. A table (`csvw:Table`) is described by its schema (`csvw:Schema`) which specifies the various columns (`csvw:Column`) it contains, as well as foreign keys (`csvw:ForeignKey`). A dialect description associated with a table provides hints to parsers on how to parse the distribution file (`csvw:delimiter`, `csvw:encoding`, etc.). The concept `csvw:TableGroup` represents a collection of datasets that share the same structure, what allows for defining the schema of these datasets for reuse. A `qb:Dataset` is associated to its structure metadata (`qb:DataStructure Definition`) as a set of measures (`qb:MeasureProperty`) organized along a group of dimensions (`qb:DimensionProperty`), together with associated metadata (`qb:AttributeProperty`). A dataset may be split into several subsets called slices (`qb:Slice`). A slice is characterized by a `qb:SliceKey` that specifies which dimensions are fixed (at least one). The `qb:concept` property allows to associate a `qb:ComponentProperty` (i.e., measure, dimension or attribute) to a concept to make its semantics explicit using domain ontologies.

The integration of these vocabularies relies on the definition in *dmo-core* of new concepts and properties (shown in orange in Fig. 1). A `dmoc:Dataset` is both a `dcat:Dataset` and a `qb:Dataset`, which allows a dataset to be

---

[5] https://www.w3.org/ns/csvw (accessed on 10th June 2022).

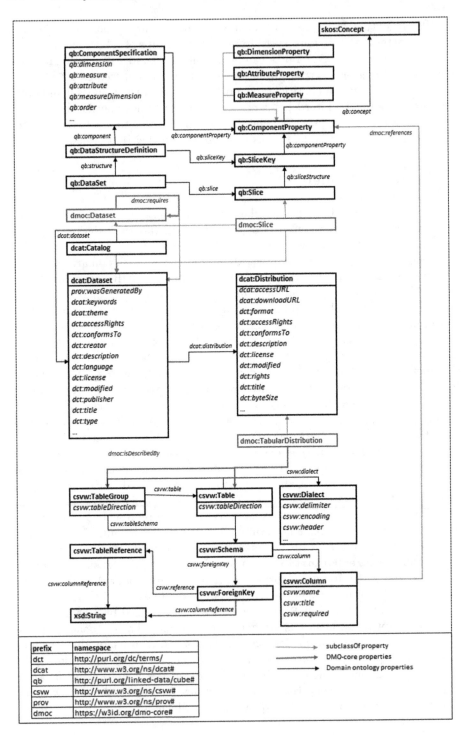

Fig. 1. The *dmo-core* model: main reused concepts and properties.

described in terms of global metadata and structure respectively. We also consider that a slice (a subset of a dataset) is a dataset, and can thus benefit from all properties of a `dmoc:Dataset`. This is why we introduce the `dmoc:Slice` concept. The concept `dmoc:TabularDistribution` is a specification of the `dcat:Distribution` to be able to describe tabular data. It is linked to a table (`csvw:Table`) or to a group of tables (`csvw:TableGroup`) with the `dmoc:isDescribedBy` property. To make the relationship between structural components (i.e., columns) and data schema components (i.e., measures and dimensions) explicit, we introduce the `dmoc:references` property between a `csvw:Column` and a `qb:ComponentProperty` (i.e., `qb:MeasureProperty` or a `qb:DimensionProperty`). For a finer semantics, we propose to link a `qb:ComponentProperty` to a concept of a domain ontology using the `qb:concept` property the range of which is `skos:Concept`. Finally, the `dmoc:requires` property aims to represent the dependency between datasets.

5 Evaluation

Several metrics, such as OntoMetrics [8], and tools such as OntOlogy Pitfall Scanner! (OOPS!) [12] can be used to evaluate ontology quality. As *dmo-core* highly relies on existing (reference) models, the quality measure here rather relies on the consistency when putting together these existing models. *dmo-core* was implemented in OWL2 and its consistency was checked thanks to different reasoners (Hermit, ELK, and Pellet) available in the Protégé[6] ontology editor. In terms of compliance to the FAIR principles, few online tools are available. One of this tools is FOOPS! [3], which takes as input an OWL ontology and generates a global FAIRness score [11]. It runs 24 different checks distributed across the 4 FAIR dimensions: 9 checks on **F** (unique, persistent and resolvable URI and version IRI, minimum descriptive metadata, namespace and prefix found in external registries); 3 checks on **A** (content negotiation, serialization in RDF, open URI protocol); 3 checks on **I** (references to pre-existing vocabularies); and 9 checks on **R** (human-readable documentation, provenance metadata, license, ontology terms properly described with labels and definitions). A score of 79% of FAIRness in FOOPS! is obtained for *dmo-core*. This score can be further improved by indexing the model in a searchable online catalog (LOV, for instance).

6 Conclusion and Future Work

This paper presented a FAIR core semantic model for descriptive and structural metadata of multidimensional tabular datasets. It was used to semantically represent several large collections of metereology datasets from the Météo-France catalog. We have now to evaluate whether the FAIRness of these metereology datasets actually helps non domain experts to reuse them.

---

[6] https://protege.stanford.edu/ (accessed on 10th June 2022).

Acknowledgement. This work is funded by the ANR (French National Research Agency) Semantics4FAIR project, contract ANR-19-DATA-0014-01.

References

1. Annane, A., Kamel, M., Trojahn, C., Aussenac-Gilles, N., Comparot, C., Baehr, C.: Towards the fairification of meteorological data: a meteorological semantic model. In: Garoufallou, E., Ovalle-Perandones, M.-A., Vlachidis, A. (eds.) Metadata and Semantic Research. pp, pp. 81–93. Springer, Cham (2022). https://doi.org/10.1007/978-3-030-98876-0_7
2. Benjelloun, O., Chen, S., Noy, N.F.: Google dataset search by the numbers. In: Proceedings of the 19th International Semantic Web Conference, pp. 667–682 (2020)
3. Garijo, D., Corcho, Ó., Poveda-Villalón, M.: Foops!: an ontology pitfall scanner for the FAIR principles. In: Seneviratne, O., Pesquita, C., Sequeda, J., Etcheverry, L. (eds.) Proceedings of the ISWC 2021 Posters, Demos and Industry Tracks: From Novel Ideas to Industrial Practice Co-located with 20th International Semantic Web Conference (ISWC 2021), CEUR Workshop Proceedings, vol. 2980. CEUR-WS.org (2021)
4. Greiner, A., Isaac, A., Iglesias, C.: Data on the web best practices. Technical report, W3C (2017). Accessed 30 Sept 2021
5. Guizzardi, G.: Ontology, Ontologies and the "I" of FAIR. Data Intell. **2**(1-2), 181–191 (2020)
6. Jacobsen, A., et al.: FAIR principles: interpretations and implementation considerations. Data Intell. **2**(1–2), 10–29 (2020)
7. Koesten, L., Simperl, E., Blount, T., Kacprzak, E., Tennison, J.: Everything you always wanted to know about a dataset: studies in data summarisation. Int. J. Hum. Comput. Stud. **135**, 102367 (2020)
8. Lantow, B.: Ontometrics: putting metrics into use for ontology evaluation. In: Proceedings of the 8th International Joint Conference on Knowledge Discovery, Knowledge Engineering and Knowledge Management - KEOD, (IC3K 2016), pp. 186–191. INSTICC, SciTePress (2016)
9. Lefort, L., Bobruk, J., Haller, A., Taylor, K., Woolf, A.: A linked sensor data cube for a 100 year homogenised daily temperature dataset. In: Proceedings of the 5th International Workshop on Semantic Sensor Networks, vol. 904, pp. 1–16 (2012)
10. Parekh, V., Gwo, J., Finin, T.W.: Ontology based semantic metadata for geoscience data. In: Arabnia, H.R. (ed.) Proceedings of the International Conference on Information and Knowledge Engineering. IKE 2004, 21–24 June 2004, Las Vegas, Nevada, USA, pp. 485–490. CSREA Press (2004)
11. Poveda-Villalón, M., Espinoza-Arias, P., Garijo, D., Corcho, O.: Coming to terms with FAIR ontologies. In: Keet, C.M., Dumontier, M. (eds.) EKAW 2020. LNCS (LNAI), vol. 12387, pp. 255–270. Springer, Cham (2020). https://doi.org/10.1007/978-3-030-61244-3_18
12. Poveda-Villalón, M., Gómez-Pérez, A., Suárez-Figueroa, M.C.: OOPS! (OntOlogy Pitfall Scanner!): an on-line tool for ontology evaluation. Int. J. Semant. Web Inf. Syst. (IJSWIS) **10**(2), 7–34 (2014)
13. Suárez-Figueroa, M.C., Gómez-Pérez, A., Fernández-López, M.: The neon methodology framework: a scenario-based methodology for ontology development. Appl. Ontol. **10**(2), 107–145 (2015)
14. van den Brink, L., et al.: Best practices for publishing, retrieving, and using spatial data on the web. Semant. Web **10**(1), 95–114 (2019)

15. Wilkinson, M., Dumontier, M., et al.: The FAIR Guiding Principles for scientific data management and stewardship. Sci. Data **3**(1), 1–9 (2016)
16. Yacoubi, N., Faron, C., Michel, F., Gandon, F., Corby, O.: A model for meteorological knowledge graphs: application to Météo-France observational data. In: 22nd International Conference on Web Engineering, ICWE 2022, Bari, Italy (2022)

Human-Centric Ontology Evaluation: Process and Tool Support

Stefani Tsaneva[1,2]([✉]), Klemens Käsznar[2], and Marta Sabou[1,2]

[1] Vienna University of Economics and Business, Vienna, Austria
{stefani.tsaneva,marta.sabou}@wu.ac.at
[2] Vienna University of Technology, Vienna, Austria
klemens.kaesznar@student.tuwien.ac.at

Abstract. As ontologies enable advanced intelligent applications, ensuring their correctness is crucial. While many quality aspects can be automatically verified, some evaluation tasks can only be solved with human intervention. Nevertheless, there is currently no generic methodology or tool support available for human-centric evaluation of ontologies. This leads to high efforts for organizing such evaluation campaigns as ontology engineers are neither guided in terms of the activities to follow nor do they benefit from tool support. To address this gap, we propose HERO - a **H**uman-Centric Ontology **E**valuation **PRO**cess, capturing all preparation, execution and follow-up activities involved in such verifications. We further propose a reference architecture of a support platform, based on HERO. We perform a case-study-centric evaluation of HERO and its reference architecture and observe a decrease in the manual effort up to 88% when ontology engineers are supported by the proposed artifacts versus a manual preparation of the evaluation.

Keywords: Ontology evaluation · Process model · Human-in-the-loop

1 Introduction

Semantic resources such as ontologies, taxonomies and knowledge graphs are increasingly used to enable an ever-growing array of intelligent systems. This raises the need of ensuring that these resources are of high quality because incorrectly represented information or controversial concepts modeled from a single viewpoint can lead to invalid application outputs and biased systems.

While several ontology issues can be automatically detected, such as logical inconsistencies and hierarchy cycles, some aspects require a human-centric evaluation. Examples are the identification of concepts not compliant with how humans think and the detection of incorrectly represented facts or controversial statements modeled from a single viewpoint [2,8]. For example, Poveda-Villalon et al. [12] identified 41 frequent ontology pitfalls, out of which 33 can be automatically detected while the remaining 8 require human judgment to be identified, e.g., P09 - Missing domain information, P14 - Misusing "owl:allValuesFrom", P15 - Using "some not" in place of "not some", or P16 - Using a primitive class in

O. Corcho et al. (Eds.): EKAW 2022, LNAI 13514, pp. 182–197, 2022.
https://doi.org/10.1007/978-3-031-17105-5_14

place of a defined one. In particular, P14 covers modeling mistakes related to universal ($\forall$) and existential ($\exists$) quantifiers stemming from the incorrect assumptions that either (1) missing information is incorrect or that (2) the universal restriction implies also the existential restriction. An example from the Pizza ontology would be a pizza *Margherita* with the two toppings *Tomato* and *Mozzarella*. Modeling these two toppings using either (1) only existential restrictions or (2) only universal restrictions, would lead to (1) pizza instances having tomato and mozzarella topping *and other toppings* or (2) pizza instances without any toppings being classified as *Margherita* pizzas.

There is a large body of literature in which ontology evaluation tasks, such as P14 above, are evaluated by humans. Indeed, a recent Systematic Mapping Study (SMS) in the field of human-centric evaluation of semantic resources identified 100 papers published on this topic in the last decade (2010-2020) [16]. A large portion of these papers (over 40%) relies on Human Computation and Crowd-sourcing (HC&C) techniques [7,15], for example, to evaluate large biomedical ontologies [9] or to ensure the quality of Linked Data as a collaborative effort between experts and the crowd [1]. In [19], we applied HC&C for supporting the evaluation of P14 through Human Intelligence Tasks (HITs) such as those shown in Fig. 1 where evaluators see a concrete entity (1, a menu item for a pizza), an axiom that models the class to which the entity belongs (2) as well as four options of (potential) errors in the axiom (3).

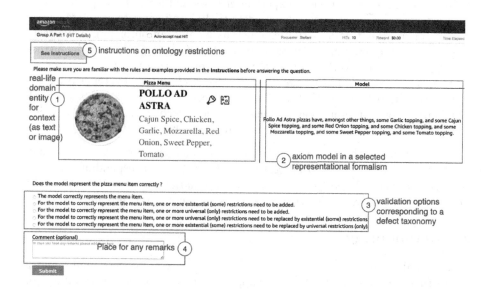

Fig. 1. HIT interface for verifying the correct use of quantifiers (from [19]).

An analysis of the 100 papers from the SMS [16] revealed two major gaps. First, there is limited understanding of the followed process by ontology engineers performing human-centric ontology evaluations: not even half of the papers outline their methodology explicitly. Some of the available methodologies are tailored to one particular evaluation aspect [11, 21] or focus on other conceptual structures than ontologies [3, 5, 17]. Second, a lack of appropriate tool support dovetails the lack of an accepted process model: indeed, less than 15% of the 100 papers mention the use of tools or libraries when preparing the evaluation.

This lack of a generalized methodology and tool support, considerably hampers the development of human-centric ontology evaluation campaigns, with each ontology engineer "reinventing the wheel" when planning such evaluations. In our own work [19], in order to prepare a human-centric ontology evaluation campaign, we could not rely on any pre-existing process or tool support and spent approximately 195 h for its realization.

In this paper, we address ontology engineers that similarly wish to prepare a human-centric ontology evaluation and aim to reduce the effort and time they need to spend on this process. To that end, we adopt a Design Science methodology [4], that leads to the following contributions in terms of concrete information artifacts and their evaluation:

- A *process model* capturing the main stages of human-centric ontology evaluation (HERO - a **H**uman-Centric Ontology **E**valuation **PRO**cess) which aims to support ontology engineers in the preparation, execution and follow-up activities of such evaluations. We focus on evaluations performed with HC&C techniques as these are currently the most frequently used. HERO was derived based on a literature review, expert interviews and a focus group.
- A *reference architecture* which supports HERO and consists of a core, which implements the general activities (such as loading an ontology), while task-specific evaluation implementations are captured as plugins and can be further extended to cater to the individual needs of evaluation tasks.
- The *evaluation of HERO and its reference architecture* by replicating the use case in [19] shows that with the support of the developed framework manual effort for preparing a human-centric ontology evaluation campaign could decrease from 30% to 88%, depending on the level of artifact reuse.

We continue with a discussion of our methodology (Sect. 2) and its main results in terms of the HERO process model (Sect. 3), a corresponding reference architecture (Sect. 4) and the use-case-based evaluation thereof (Sect. 5). Lastly, we present related work (Sect. 6) and discuss concluding remarks in Sect. 7.

2   Methodology

As our goal is to establish two information artifacts (a process model and a supporting reference architecture), we apply the *Design Science methodology for information systems research* [4] realized in three steps as illustrated in Fig. 2: Step 1 and Step 2 cover the *development phase* of the two artifacts, while Step 3 represents the *evaluation phase* based on a concrete case study. In Step 1, we incorporate knowledge from existing literature into the design of the artifacts (*rigor cycle*) while also involving key stakeholders in need of such artifacts (*relevance cycle*). The details of each methodological step are described next:

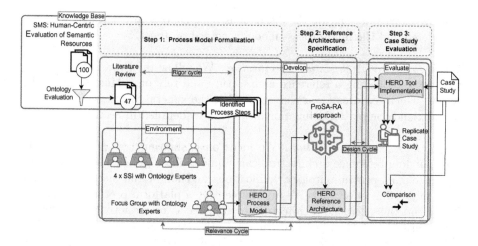

Fig. 2. Design-Science-based methodology.

Step 1: Process Model Formalization relies on three diverse methods for deriving the HERO process model. First, we review existing literature discussing ontology evaluation relying on human involvement. To that end, we leverage ongoing work on a *Systematic Mapping Study (SMS)* about human-centric evaluation of semantic resources [16]. From the set of 100 papers collected by the SMS, we identified 47 papers that discuss *ontology* evaluation and review them to identify typical activities followed when performing human-centric ontology evaluation and the tools used in that process. As a result, we collect a set of activities to be included in HERO and group them into three stages: preparation, execution and follow-up. We structure the data collected from these 47 papers in a Knowledge Graph, published at our git repository[1], making it available to other researchers.

Second, we perform a set of *semi-structured interviews (SSI) with experts in Ontology Engineering* to uncover missing aspects not described in the papers from the literature, discuss order of activities and required tools. Interviewees were selected from the Vienna University of Technology and included a senior researcher, a Ph.D. student, a graduate student, and a master's student, each conducting work in the area of human-centric ontology evaluation. During the interviews, a set of activities, part of human-centric ontology evaluations, are identified from the perspective of each expert as well as tools they used when conducting past evaluations. The interviews aim at strengthening and supporting the findings from the literature corpus and both approaches are designed independently to ensure that the experts are not biased and their personal views on the process are captured. More details on the SSI and a comparison of the steps found in literature vs. those identified during the SSI can be found in [6].

Third, we conduct a *focus group with the experts* that participated in the interviews to combine the literature analysis results with the insights gathered

---

[1] github.com/k-klemens/hc-ov-process-models/tree/main/slr.

from the interviews. During the discussions open aspects are clarified, activities are ordered and the final process model is agreed on (Sect. 3).

Step 2: Reference Architecture Specification. We follow the ProSA-RA [10] approach for establishing a reference architecture and rely on a Microkernel architecture, which features (1) a core, capturing the general logic and (2) plugins, which extend the platform functionalities [14], as detailed in Sect. 4.

Step 3: Case Study Evaluation. We focus on evaluating how HERO and the corresponding reference architecture can support ontology engineers when conducting human-centric ontology verifications and follow the methodology proposed by Wohlin et al. [22]. We first implement a platform prototype based on HERO and the reference architecture to support as many activities in the use case described in [19] as possible. Subsequently, we compare the effort required to prepare the evaluation campaign with HERO and tool support against the effort of manually creating the evaluation in the original use case (Sect. 5).

3   HERO - A Human-Centric Ontology Evaluation Process

HERO is a process model for human-centric ontology evaluation, targeted toward micro-tasking environments such as crowdsourcing platforms and focusing on batch-style evaluations. At a high-level the process and its activities can be structured into the stages of *preparation, execution* and *follow-up*, as detailed in the next sections and exemplified by the use case from [19] introduced in Sect. 1. The *preparation* stage consists of the design and definition of the evaluation, while during the execution stage the evaluation is conducted, followed by the *follow-up* stage where the evaluation data is collected and analyzed. Note that HERO aims at being broadly applicable and as a result includes activities that might not be needed in every human-centric ontology evaluation.

3.1   Preparation Stage

The activities part of the HERO preparation stage are visualized in Fig. 3. A full black circle indicates the start of the process while a black circle surrounded by a white circle represents the end of the process. Parallel activities are situated between two vertical lines, while connected activities are placed into activity groups (e.g., "Task design").

As a starting point, the *ontology to be verified needs to be loaded* (1 in Fig. 3) and to get an overview of the ontology, *standard metrics and quality aspects should be inspected* (2; e.g., in Protégé, among other things, the number of axioms, classes and data properties can be explored). A crucial preparation activity is the *specification of the aspect for evaluation* (3) and the overall goal of the verification. In the specified use case, the correct usage of ontology restrictions is the aspect to be verified. *Specifying the evaluation environment* (4) is the

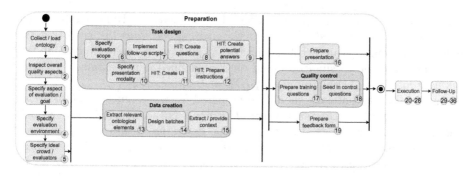

Fig. 3. HERO preparation stage.

next activity which refers to deciding on a crowdsourcing platform (e.g., in [19] Amazon Mechanical Turk (mTurk, mturk.com) was used) or another medium (e.g., games with a purpose, pen&paper, custom interface) that enables conducting the verification. In addition, the *ideal crowd's characteristics* (5; e.g., demographics, expertise in the domain) should be specified at an early stage of the process as these could have an impact on the verification task design. Nevertheless, special consideration should be taken with regard to avoiding creating a potential bias through crowd selection. In the evaluation performed in [19] we decided on an internal student crowd rather than a layman crowd since general modeling understanding of the evaluators and a more controlled environment were prerequisites. We further asked the students to complete a self-assessment to gain a better overview of their background in several areas (e.g., English skills, modeling experience). Evaluators' skills were also tested by implementing a qualification test, which offered an objective assessment of the evaluators' prior knowledge of the quantifiers. Further details on these assets can be found in [18].

Next, the verification *task design* follows. Several activities are to be expected, which have no particular order, since the task design process is iterative. *Specifying the evaluation scope* (6) can include specifying what subset of the ontology to show to the evaluators. In [19] all restrictions on a single relation are grouped together forming ontology restriction axioms that fully describe the specific relation and can be evaluated independently from the rest of the axioms. However, to verify a subclass relation it might be sufficient to only present the ontology triple, while for judging the relevance of a concept, more ontological elements might be needed to ensure a correct judgment.

As the task design might impose the structure of the final data and implications for analysis arise, *follow-up scripts* (7) for data processing can be implemented. In [19] analysis scripts are implemented in R (r-project.org) and tested in the preparation stage to avoid unexpected issues at the final process stage. The *specification of presentation modality* (10) implies deciding on the representation of ontology elements that evaluators will see. In the specified use case, we considered 3 representations- two plain text axiom translations, proposed by the authors of [13] and [20] as alternatives to showing OWL to novice ontology engineers, and the graphical representation VOWL. Next, the Human Intelligence Tasks (HITs) are designed which includes *creating questions* (8), such that the required information can be collected, *creating potential answers* (9) (e.g., in [19] we created answer possibilities based on a predefined defect catalog), *creating the user interface* (11) and *preparing HIT instructions* (12).

In parallel, the ontology should be prepared for the evaluation (*data preparation*) by *extracting relevant ontological elements* (13), which in [19] we accomplished using Apache Jena (jena.apache.org), *designing batches* (14) so that relevant tasks are grouped together, and *extracting context* (15) to be presented to the evaluators (e.g., in [19] pizza menu items are manually created) to be provided to the evaluator.

In some evaluations it is beneficial to *create a presentation* (16) to inform the evaluators what the verification is about and what their assignment is (e.g., expert evaluation). In [19] we prepared a presentation for the student crowd, which included the main goals of the evaluation, instructions and tips on using mTurk, and organizational aspects.

Another important activity group is the *quality control*. One approach is to *prepare training questions* (17) to be completed by the evaluators prior to the actual verification, similar to the qualification tests which can ensure that the crowd acquires a particular skill. For the evaluation performed in [19] several tutorial questions were available to ensure the students are familiar with the mTurk interface and tasks format prior to the actual verifications and as aforementioned qualification tests were also completed by the participants. Another option is to *seed in control questions* (18) based on a (partial) gold standard without the evaluators' knowledge, which allows for assessing the intention or trustworthiness of the workers later on when the judgments are aggregated (e.g., filtering out spammers or malicious workers). In the described use case [19] it was not necessary to seed in control questions since a gold standard was available and the evaluation only had experimental aims.

Lastly, a *feedback form preparation* (19) might be especially useful in evaluation cases, where the verification should be repeated and the process should be improved based on comments from the evaluators. In [19] we collect feedback to analyze the students' experience when performing the verifications and outline confusing aspects that could be improved.

3.2   Execution Stage

Once the preparatory activities are completed, the verification tasks need to be performed following the activities depicted in Fig. 4. First, the *HIT templates are populated* (20 in Fig. 4) with data. At this point, the tasks are not yet publicly accessible and can be refined if needed before the evaluators can start working on them. Next, the *HITs are published* (21) and if needed *a presentation is shown* (22) to ensure the evaluators are familiar with the verification tasks.

To ensure high-quality judgments, *qualification tests* (23) can be made a requirement for working on a HIT. In [19], a high score of the developed qualification test for the crowd's skills in ontology modeling was not a prerequisite

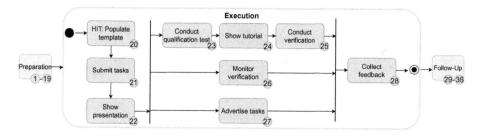

Fig. 4. HERO execution stage.

since the performed experiments aimed at analyzing how prior knowledge affects the verification results. Next, *the tutorial tasks are shown* (24), which consists in showing the previously prepared training questions. Afterward, the *verification is conducted* (25), which is the main activity in this stage where the verification tasks are completed by the evaluators. Typically the evaluation environment (e.g., mTurk) is responsible for showing open batches of questions and collecting the answers from the evaluators.

In parallel to the previously described activities, the *verification should be monitored* (26), which ensures potential problems are identified and corrected early on. To that end, crowdsourcing platforms typically provide management interfaces that can be used. In some cases, it might be required to stop the process and go back to a previous activity for revision. During the evaluations performed in [19] a Zoom Meeting was active, where evaluators could ask questions and technical problems were solved. Another parallel activity is the *advertisement of tasks* (27), which can be achieved through different means such as newsletters, web pages, or any other communication means. This activity is of particular importance if not enough evaluators are engaging in the tasks.

Finally, *feedback is collected* (28) from the evaluators using the prepared feedback form, so that potential problems with the workflow can be identified.

3.3   Follow-Up Stage

Follow-up activities conclude the ontology evaluation process as depicted in Fig. 5. Initially, the *crowdsourced data is to be collected* (29 in Fig. 5) and *preprocessed (30)* to be compatible with the prepared data analysis scripts. To gain an overview of the collected judgments, *data quality statistics are calculated,* which can include (but is not limited to) *calculating trustworthiness* (31) based on the control questions and other measures provided by the evaluation environment and *calculating inter-rater agreement* (32).

In micro-tasking environments typically redundant judgments are collected for each task. To obtain a conclusion on a task these answers need to be aggregated (33; e.g., using majority voting as in [19]). Afterward, the *data needs to be analyzed* (34) in order to obtain the final results of the verification.

Once a final set of results is obtained through analysis, this can be used to *improve the verified ontology* (35). This activity is tightly linked to the goal

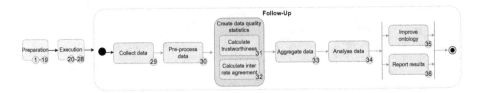

Fig. 5. HERO follow-up stage.

specified during the preparation and depending on it, the results can be used to improve certain aspects of the ontology. In [19] the participants were provided with their verification scores to enable the learning process and results were analyzed to test the experiment hypotheses. At the same time, the *results are to be reported* (36).

4   HERO Reference Architecture

The HERO process provides an in-depth understanding of the activities typically performed during human-centric ontology evaluation, and as such enables the design of a *reference architecture* that can be used as a basis for creating platforms that (partially) automate the activities of such evaluation processes.

HERO contains both activities that are relevant for a wide range of evaluation campaigns (e.g., loading and inspecting the ontology) as well as activities that are specific to certain evaluations (e.g., task design). To that end, we relied on a Microkernel Architecture which features (1) a core, where the general logic is captured and (2) the plugins, which extend the platform functionalities [14] (Fig. 6). Accordingly, the general functionalities of the platform are included as core components (i.e., an *Ontology Loader, Ontology Metrics, Data Provider, Triple Store, Verification Task Creator, Quality Control, Crowdsourcing Manager, Meta Data Store* and *Data Processor*), while plugins allow for customization to specific use cases. For instance, different *Context Provider* plugins can be developed to extract relevant context to be presented to the evaluators in the HITs and a separate *Crowdsourcing Connector* is needed for each crowdsourcing platform. Further information on the reference architecture (i.e., crosscutting, deployment and run-time viewpoints) can be found in [6].

5   Case-Study-Based Evaluation

The evaluation of the created artifacts investigates: *To what extent can the HERO process model and the corresponding reference architecture support the preparation of a human-centric ontology evaluation (i.e., the HERO preparation stage)?* We focus the evaluation on the *preparation stage of HERO* as it is the most effort intensive and can be most reliably replicated. Our evaluation goal translates into the following sub-questions:

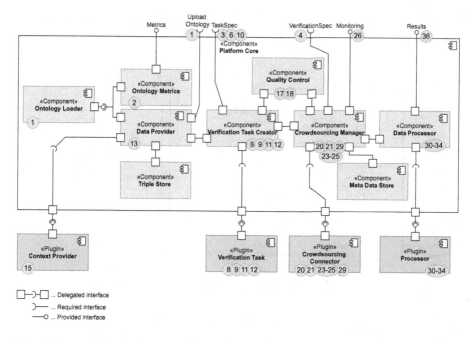

Fig. 6. Source code viewpoint of the HERO reference architecture including numbers of the connected process activities (see also Figs. 3, 4 and 5).

– *RQ1: Can the HERO process model be used to better structure the activities of a concrete evaluation campaign?*
– *RQ2: Is it feasible to implement a supporting platform based on the reference architecture?*
– *RQ3: How many preparation activities can be automated by a HERO-based platform?*
– *RQ4: What is the effort reduction when using the platform as opposed to a manual preparation process?*

To answer these research questions, we adopt a Case Study methodology [22] consisting of replicating the use case described in [19] by making use of the artifacts we developed. We started by representing the activities we followed during the preparation of the evaluation campaign from the use case in terms of the HERO process model in Fig. 7. We found that HERO can contribute to more clearly structuring how and through which activities the preparation of the evaluation campaign was performed (RQ1). It can also highlight potential weaknesses, for example, that the original preparation did not cover three activities: inspecting the ontology quality, designing batches and seeding control questions (which could be beneficial additions). Subsequently, as per RQ2, we used the reference architecture as a basis to implement a prototype platform to support the use case activities (Sect. 5.1). The prototype platform allowed the tool-supported replication of the use case and enabled a comparison with the effort spent during the manual execution of the use case (RQ3, RQ4, Sect. 5.2).

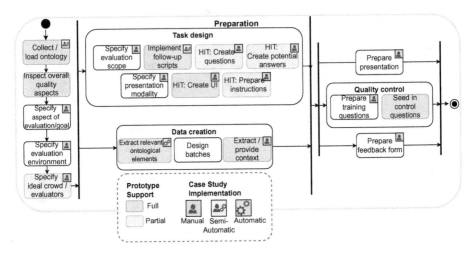

Fig. 7. Structured representation of the original use case (from [19]) in terms of the HERO activities (with indication of whether they were performed manually, automatically or semi-automatically). Indication of which of the activities of the process can be fully/partially supported by the HERO-based prototype platform.

5.1   Feasibility Study: Prototype Implementation

Following Fig. 6 of the reference architecture, we implement all core components[2] except the *Ontology metrics* (since this component is not required for replication of the use case [19]). To replicate the use case from [19], we develop several customized plugins[3], as follows:

- *RestrictionVerificationPlugin (VT)*, responsible for defining how the universal and existential quantification axioms are extracted from a given ontology. Further, we specify an HTML template and a method on how to extract values from the ontology for each variable in a template to define the GUI of the HITs. By using a configuration property the axioms can be rendered in the representational formalism proposed in [13] and [20].
- *PizzaMenuContextProviderPlugin (CP)*, responsible for creating a restaurant-menu-styled-item for each pizza ontology axiom.
- *AMTCrowdsourcingConnector (CC)*, responsible for publishing tasks on mTruk, retrieving the current status of the published verification and also obtaining the raw results from the platform.

---

[2] github.com/k-klemens/hc-ov-core.
[3] github.com/k-klemens/hc-ov-pizza-verification-plugins, ../hc-ov-amt-connector.

We conclude that the reference architecture was sufficiently detailed to make the implementation of a concrete supporting platform *feasible (RQ2)*. The implementation of the platform core took 55 h while the use-case-specific plugins required 28.5 h. We expect that similar implementation efforts will be required for other technology stacks or evaluation use cases.

5.2   Evaluation Results

Automated Activities (RQ3). As color-coded in Fig. 7, with the implemented platform prototype we can fully support 7 out of 19 (37%) and partially support 4 out of 19 (21%) of the HERO preparation activities. In the context of our use case [19], which only covered 16 activities, over 55% of the conducted preparation activities can be (fully or partially) supported by the platform. Activities, which are not supported by the platform (e.g. "Specify evaluation scope") require human decisions or are not expected to be reusable once automated. Further, going beyond preparation activities, the platform and the implemented plugins allow publishing, monitoring and retrieving the results of created HITs on mTurk (although these activities are not subject to the current evaluation).

Reduction of Time Effort (RQ4). Our baseline is the time effort that was spent to prepare the ontology evaluation campaign in the original use case [19], that is 195 h. As part of this case study, we replicated the use case with the tool support based on HERO. The total effort spent on this replication amounts to 137 h (Fig. 8) and includes: specifying the reference architecture (48 h), implementing the platform core (55 h), implementing specific plugins (28.5 h), and miscellaneous activities, e.g., meetings (5 h).

Comparisons of these two effort categories can be performed in two ways. First, assuming that this replication case study is a one-of-activity, the effort for the preparation stage of the use case could be reduced by 30% by adopting the principled HERO-based approach and relying on the corresponding tool support. Second, our aim is that the artifacts created so far can be re-used in follow-up projects. In that case, assuming that in this case study we would have reused the reference architecture and core platform implementations, the effort for replication consists only in the adaptation of the plugins (28.5), thus leading to an effort reduction of 88%. More details are available in [6].

Improved Aspects. Besides time effort reduction, the platform offers *centralized orchestration and storage* by implementing end-to-end process support for human-centric ontology verification. It also allows for *extensibility and reusability* via the plugin architecture. Once a plugin is implemented, it can be reused for future verifications, thus, overall implementation efforts are expected to be reduced as the availability of plugins grows. Since the platform allows for the automation of manual activities (e.g., the extraction of context), *data scalability* will be ensured, especially for larger ontologies.

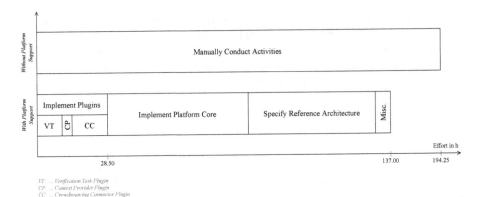

Fig. 8. Comparison of efforts between the use case evaluation approach *without platform support* and *with platform support*.

6   Related Work

Several works have already explored the conceptualization of a verification process model with a human-in-the-loop for evaluations of Semantic Web artifacts. In [5] the authors propose a high-level methodology for the crowdsourcing-based assessment of the quality of Linked Data resources, partly supported by the TripleCheckMate tool. However the methodology and tool are very much geared towards *assessing linked data triples*. A new method for *task-based ontology evaluation* is proposed by the authors of [11]. The presented methodology is tailored towards application ontologies and how fitted they are for a certain application task. In [21] the authors describe *a plugin for Protégé* supporting ontology modelers in the Ontology Engineering process, by outsourcing a set of human-centric tasks to games with a purpose or a crowdsourcing platform. However, this approach is dependent of the Protégé editor, which reduces its reusability.

Additionally to the Semantic Web research, in the Software Engineering domain, we are aware of related work such as (1) an approach for the *verification of Enhanced Entity-Relationship diagrams* based on textual requirement specifications relying on HC&C techniques [17] and (2) a process and framework for the human-centirc *validation of OntoUML (*ontouml.org*) models* [3].

Only two [11, 21] of the process models above focus on the evaluation of ontologies. Most of the process models are lacking key details and have been derived ad-hoc as opposed to following a principled approach. Therefore, a need arises for a human-centric ontology evaluation method with more details and which is derived in a methodologically principled way.

7   Summary and Future Work

While human-centric ontology evaluation is often performed in order to verify ontology quality aspects that cannot be identified automatically, this area currently lacks detailed methodologies and suitable tool support. In this paper we address this gap by providing (1) HERO- a detailed process for preparing and conducting human-centric ontology evaluation; (2) a reference architecture that supports this process; (3) a case-study-based evaluation exemplifying the use and benefits of these artifacts during the reproduction of a concrete use case. The case study indicates that, when supported by the HERO process and platform, ontology experts require, depending on the level of the artifact reuse, between 30% and 88% less time to prepare an ontology verification compared to a manual setting. As more plugins become available for reuse, we expect further reductions of effort, especially in use cases dealing with large ontologies. All artifacts were derived in a methodologically principled way by covering all three Design Science cycles and shared in the GitHub repository together with additional information.

In *future work*, we aim to address some of the current *limitations*. While we carefully followed a Design Science methodology, the core cycle of this method was only performed once. Therefore, we wish to conduct more design-evaluation cycles to further improve the current artifacts. Along the *evaluation* axis, the reference architecture was only indirectly evaluated through the case study's evaluation. A more sophisticated evaluation approach could involve interviews with domain experts and software architects. The case study focused on a use case with a small ontology thus giving only partial insights into efficiency gains, especially for larger ontologies. We plan a number of follow-up replication studies with larger ontologies and different verification problems to further test our artifacts. Along the *design* axis, further evaluations as described above will lead to iterative improvements of the artifacts such as (1) formalizing the HERO process using standards such as Business Process Model and Notation (BPMN) in order to create a richer model including also information about roles and activity results; (2) extending the implementation of the platform core and creating additional plugins for other types of ontology verification problems as well.

Acknowledgments. We thank all interview and focus group participants. This work was supported by the FWF HOnEst project (V 754-N).

References

1. Acosta, M., Zaveri, A., Simperl, E., Kontokostas, D., Auer, S., Lehmann, J.: Crowd-sourcing linked data quality assessment. In: ISWC, pp. 260–276 (2013)
2. Erez, E.S., Zhitomirsky-Geffet, M., Bar-Ilan, J.: Subjective vs. objective evaluation of ontological statements with crowdsourcing. In: Proceedings of the Association for Information Science and Technology, vol. 52, no. 1, pp. 1–4 (2015)
3. Fumagalli, Mattia, Sales, Tiago Prince, Guizzardi, Giancarlo: Mind the Gap!: learning missing constraints from annotated conceptual model simulations. In: Serral, Estefanía, Stirna, Janis, Ralyté, Jolita, Grabis, J.ānis (eds.) PoEM 2021. LNBIP, vol. 432, pp. 64–79. Springer, Cham (2021). https://doi.org/10.1007/978-3-030-91279-6_5

4. Hevner, A.R., March, S.T., Park, J., Ram, S.: Design science in information systems research. MIS Q. 75–105 (2004)
5. Kontokostas, D., Zaveri, A., Auer, S., Lehmann, J.: TriplecheckMate: a tool for crowdsourcing the quality assessment of linked data. Commun. Comput. Inf. Sci. **394**, 265–272 (2013)
6. Käsznar, K.: A process and tool support for human-centred ontology verification. Master's thesis, Technische Universität Wien (2022). https://repositum.tuwien.at/handle/20.500.12708/20577
7. Law, E., Ahn, L.V.: Human computation. Synth. Lect. Artif. Intell. Mach. Learn. **5**(3), 1–121 (2011)
8. McDaniel, M., Storey, V.C.: Evaluating domain ontologies: clarification, classification, and challenges. ACM Comput. Surv. (CSUR) **52**(4), 1–44 (2019)
9. Mortensen, J.M., et al.: Using the wisdom of the crowds to find critical errors in biomedical ontologies: a study of SNOMED CT. J. Am. Med. Inform. Assoc. **22**(3), 640–648 (2015)
10. Nakagawa, E.Y., Guessi, M., Maldonado, J.C., Feitosa, D., Oquendo, F.: Consolidating a process for the design, representation, and evaluation of reference architectures. In: 2014 IEEE/IFIP Conf. on Software Architecture,pp. 143–152 (2014)
11. Pittet, P., Barthélémy, J.: Exploiting users' feedbacks: towards a task-based evaluation of application ontologies throughout their lifecycle. In: IC3K 2015 - Proceedings of the 7th International Joint Conference on Knowledge Discovery, Knowledge Engineering and Knowledge Management, vol. 2, pp. 263–268 (2015)
12. Poveda-Villalón, M., Gómez-Pérez, A., Suárez-Figueroa, M.C.: Oops!(ontology pitfall scanner!): an on-line tool for ontology evaluation. Int. J. Semant. Web Inf. Syst. (IJSWIS) **10**(2), 7–34 (2014)
13. Rector, A., et al.: OWL Pizzas: practical experience of teaching OWL-DL: common errors & common patterns. In: Motta, Enrico, Shadbolt, Nigel R., Stutt, Arthur, Gibbins, Nick (eds.) EKAW 2004. LNCS (LNAI), vol. 3257, pp. 63–81. Springer, Heidelberg (2004). https://doi.org/10.1007/978-3-540-30202-5_5
14. Richards, M.: Software Architecture Patterns, vol. 4. O'Reilly Media, Incorporated 1005 Gravenstein Highway North, Sebastopol, CA (2015)
15. Sabou, M., Aroyo, L., Bontcheva, K., Bozzon, A., Qarout, R.: Semantic web and human computation: the status of an emerging field. Semant. Web J. **9**(3), 291–302 (2018)
16. Sabou, M., Fernandez, M., Poveda-Villalón, M., Suárez-Figueroa, M.C., Tsaneva, S.: Human-centric evaluation of semantic resources: a systematic mapping study, In preparation
17. Sabou, M., Winkler, D., Penzerstadler, P., Biffl, S.: Verifying conceptual domain models with human computation: A case study in software engineering. In: Proceedings of the AAAI Conference on Human Computing and Crowdsourcing, vol. 6, pp. 164–173 (2018)
18. Tsaneva, S.: Human-Centric Ontology Evaluation. Master's thesis, Technische Universität Wien (2021). https://repositum.tuwien.at/handle/20.500.12708/17249
19. Tsaneva, S., Sabou, M.: A human computation approach for ontology restrictions verification. In: Proceedings of the AAAI Conf. on Human Computation and Crowdsourcing (2021). www.humancomputation.com/2021/assets/wips_demos/HCOMP_2021_paper_90.pdf
20. Warren, P., Mulholland, P., Collins, T., Motta, E.: Improving comprehension of knowledge representation languages: a case study with description logics. Int. J. Hum.-Comput. Stud. **122**, 145–167 (2019)

21. Wohlgenannt, G., Sabou, M., Hanika, F.: Crowd-based ontology engineering with the uComp Protégé plugin. Semant. Web **7**(4), 379–398 (2016)
22. Wohlin, C., Runeson, P., Höst, M., Ohlsson, M.C., Regnell, B., Wesslén, A.: Experimentation in Software Engineering. Springer, Heidelberg (2012). https://doi.org/10.1007/978-3-642-29044-2

Position Papers

Towards Pragmatic Explanations
for Domain Ontologies

Elena Romanenko[1(✉)] , Diego Calvanese[1,2] , and Giancarlo Guizzardi[1,3]

[1] Free University of Bozen-Bolzano, 39100 Bolzano, Italy
{eromanenko,giancarlo.guizzardi}@unibz.it, calvanese@inf.unibz.it
[2] Umeå University, 90187 Umeå, Sweden
[3] University of Twente, 7500 Enschede, The Netherlands

Abstract. Ontologies have gained popularity in a wide range of research fields, in the domains where possible interpretations of terms have to be narrowed and there is a need for explicit inter-relations of concepts. Although reusability has always been claimed as one of the main characteristics of ontologies, it has been shown that reusing domain ontologies is not a common practice. Perhaps this is due to the fact that despite a large number of works towards complexity management of ontologies, popular systems do not incorporate enough functionality for ontology explanation. We analyse the state of the art and substantiate a minimal functionality that the system should provide in order to make domain ontologies better understandable for their users.

Keywords: Ontology Explanation · Pragmatic Explanation · Ontology engineering

1 Introduction

In the last few years, eXplainable AI has become a subject of intense study. Many efforts have been carried out towards an interpretation of how deep neural networks produce their results, and why we can trust them. At the same time, there is a common belief that ontologies (and other symbolic artefacts) do not suffer from the same problems and that current systems provide enough support to understand and reuse already existing artefacts. However, it has been shown that reusing ontologies is not a consolidated practice [5].

We claim that one reason for this could be the lack of support from software tools to make ontologies better understandable. Thus, the goal of this paper is to answer the following open questions: *(1)* What is an explanation in case of ontologies, and why we may be interested in it? *(2)* What is the minimal functionality a system should provide in order to make ontologies understandable?

The remainder of the paper is organized as follows. Section 2 determines relationships between ontologies and explanations and provides context for the rest of the paper. Section 3 describes a pragmatic approach to studying explanations of ontologies. Section 4 substantiates the functions a system should provide in

O. Corcho et al. (Eds.): EKAW 2022, LNAI 13514, pp. 201–208, 2022.
https://doi.org/10.1007/978-3-031-17105-5_15

order to provide explainability. Finally, Sect. 5 concludes the paper and points to open issues requiring further investigation.

2   Ontology and Explanation

2.1   Ontology as an Explanation

Since when they first appeared as "formal and explicit specification of shared conceptualizations" [9] in the beginning of the 90's, ontologies have gained popularity in a wide range of research fields. These range from knowledge engineering and knowledge representation to database design and information integration, in all those domains where possible interpretations of terms have to be constrained and there is a need for explicit inter-relations of concepts.

Fonseca and Martin [6] argue that ontologies should lead to explanation and understanding of the domain. Also, Cao [4] suggested an *ontological approach to explanation*: "whenever we have something important but difficult to understand, we should focus our attention on finding what the primary entities are in the domain under investigation. Discovering these entities and their intrinsic and structural properties [...] is the real work of science". Garcia and Vivacqua [8] concluded that "ontology explanation can be a valuable tool for semantic validation", where the latter term is defined as "human understanding of and agreement with the ontological representation".

Thus, it is generally accepted that *an ontology may serve as an explanation within the domain of interest*. And, according to this approach, ontology usage should result in an understanding of the domain (see [13, 15] and [20, p.4]). However, does this imply that an ontology itself is such a clear, self-explained, and understandable artefact? Unfortunately, recent reports have shown, that despite existing well-accepted methods, "building an ontology is sometimes equated to an art" [8], and, even when ratified by experts, it may lead to unexpected results in trials. This brings us to the idea that *an ontology as an artefact also needs to be explained*.

2.2   Aspects of Ontology Explanation

The history of scientific explanations can be traced far back into antiquity, with numerous philosophical discussions in the second half of the 20th century (see [19]). This subsection is not intended to be an overview of existing theories in the philosophy of science but instead an attempt to find those that are applicable in the case of ontology explanation.

Hempel [12, p.334] formulated that "scientific explanation may be regarded as an answer to a why-question". According to his covering-law model, a scientific explanation takes the form of a syllogism consisting of a law, a statement of facts making up the initial conditions, and a statement of the event which occurred. He suggested two basic types of explanations, deductive-nomological and inductive-statistical, differing in how the conclusions are drawn. However, this approach resembles data provenance techniques or two ways of reasoning, where the ontology plays the role of a 'body of (domain) laws' within the domain of interest, while our goal is to explain the ontology itself.

It is worth noting that explanations differ depending on the traditions within the branch of science. For example, Baron [1] claims that psychology does not offer scientific explanations but rather "follows certain templates (schemata), basic forms with details to be filled in". In contrast to this approach, mathematicians explain their conclusions using formal proofs, e.g., in first-order logic.

Taking this into account, Weber et al. [20] suggest *a pragmatic approach to studying scientific explanation*. Here 'pragmatic' refers to an 'instrumentalist' approach towards constructing explanations, so as to consider a 'toolbox' that would help actors to reach their goals, instead of "describing the 'essence' of explanation or understanding" [20, p.33].

In order to apply this pragmatic approach to studying explanations, the authors suggest committing to three principles [20]:

1. *Make context-dependent normative claims and argue for them.* Here 'normative' refers to looking at how scientists actually explain in real-life scenarios, while 'context-dependent' means that we should work only in a certain discipline and within certain research traditions.
2. *Make context-dependent descriptive claims and argue for them.* The authors claim that when trying to 'describe' the existing practices, there is no need for a large sample of scientists to be interviewed or a large number of scientific writings to be analysed.
3. *Take into account the epistemic interests*[1] when trying to make context-dependent normative or descriptive claims about explanations.

Also, while applying the pragmatic approach, the following general considerations about explanations need to be taken into account. Miller [16] argues that there is a need to distinguish between "explanation" as *(1)* a cognitive process, including abductive reasoning and reduction to filling gaps, *(2)* a product, that can come in different forms[2], and *(3)* a social process of interaction.

The first process is what Lipton [15] refers to as closing the gap between knowledge and understanding. Since "the explained phenomenon is said to be reduced to the explaining phenomenon" [7], and the cognitive process of understanding and reduction happens in the human mind, it follows that the explanation of an ontology is *a user-centred process*[3].

According to Horne et al. [13], "an explanation is a statement that satisfies the request for information". However, as has been noted by Miller [16], this request for information often forms an interaction process. Lipton [15] suggested *a why regress* as one of the features of explanation, because "... explanation

---

[1] An epistemic interest is a reason scientists have for asking explanation-seeking questions.

[2] We come back to the discussion about forms of explanation later.

[3] Henceforward, by 'user' we mean a domain expert or a developer or simply anyone interested in getting an ontology explanation from the computer system.

can bring us to understand why what is explained is so even though we do not understand why the explanation itself is so".

Finally, *explanations are contrastive* (see [13,15,16]). In other words, people rarely ask "why P" but are more interested in "why P rather than Q". Horne et al. [13] even stated that "a contrast class of similar but non-observed states" should be incorporated within the structure of the explanation together with the explanandum (the observation to be explained) and "a request for information that differentiates the occurrence of the explanandum from the nonoccurrence of its contrast class".

3   A Pragmatic Approach to Ontology Explanation

As has been described before, a pragmatic approach to studying explanations suggests us to consider how scientists explain ontologies in real-life scenarios.

Garcia and Vivacqua [8] claimed that "the explanation played a fundamental role in the identification of problems" with the ontology. The main goal was to make the evaluation process more productive, and to help ontology users and developers to understand and audit the ontology. The authors reported about *visual inspection* as well as *text explanations* generated in a story-like format. They also noticed that using domain cases, i.e., *examples*, during the ontology development is a good way to elicit conflicts between experts' opinions [8].

A quite similar example of a storytelling approach was presented by Braga and Almeida [2], who used natural language narratives with the concepts that appear in the conceptual model for assessing and 'testing' the correctness of those models. That work was an extension of [3], which employed visual simulation for revealing the semantics of the ontology.

Horridge et al. [14] suggested a toolkit for working with *justifications for entailments* in ontologies, presented to the user as *a list of logical axioms*. Justifications are considered as a specific type of explanation, where a minimal sufficient subset of the ontology is selected to hold the entailment [18].

Nevertheless, despite all the efforts suggested in [8] towards ontology explanation, the first version of the developed artefact exhibited some critical problems. Taking into account how knowledge acquisition sessions were organised, we suppose that the explanation-as-a-process (both as a cognitive process as well as a social process) was properly supported. However, in contrast to the approach in [14], the authors were more focused on changing *the form of explanation* (visual and textual), rather than changing the explanation-as-a-product itself.

Overton [17] suggested considering five categories and relations between them (see Fig. 1). According to him, an explanation consists of an *explain-relation*, an *explanan*, i.e., an explaining category, and an *explanandum*, i.e., an explained category. Unlike Hempel, he does not restrict explanans and explanandum to propositions but considers a broader view. He argues that explanations of phenomena at one level could be relative to and refer to another level, so explanations between two levels should refer to all intermediate levels.

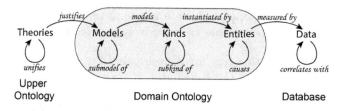

Fig. 1. Overton's five categories with some relations between them.

Following this approach, we can consider upper-level ontologies, such as DOLCE[4], or UFO [11], in the category of 'theories', and domain ontologies or ontology-driven conceptual models based on them at the lower levels. Thus, we can partially explain a domain ontology by *(1)* grounding it on an upper-level one, *(2)* increasing its comprehensibility and semantic transparency by leveraging on foundational techniques for complexity management[5], or *(3)* justifying it by providing data evidence. Aspects (1) and (3) are exemplified in Sect. 4.

The last question is whether an ontology as a way to provide explanation to an ontology is reasonable and valid. Actually, this form of explanations is known as *self-evidencing explanations*: "... explanations where what is explained provides an essential part of our reason for believing that the explanation itself is correct" [15]. So while an upper-level ontology partially explains (grounds) a domain ontology, the domain ontology justifies the upper-level one.

Therefore, when providing an ontology explanation, the following ideas need to be taken into account. First, the ultimate goal of the user is rarely on understanding the ontology per se, but rather on understanding it such that its quality can be assessed, and so that it can be safely reused and integrated with other ontologies. Second, the explanation of a domain ontology should be provided at different levels, by explicitly referencing (grounding it on) a foundational ontology, by increasing its comprehensibility and transparency via complexity management, and also justifying it with entailments or examples from real data. Third, the explanations can be given in different forms (visual, text, logic), but the form of explanation can be considered as a user's preference of presenting an explanation-as-a-product. Finally, the system should be ready for explanation-as-a-process in the sense that a user may ask to explain the explanation.

4   Functionality of an Ontology Explanation System

An Explanation System should provide an opportunity for a potential regress of explanation, i.e., the possibility to explain the explanation. In case of ontology explanation, this interaction may happen between the system and the user, so it is a human-computer interaction process, which is dependent on the user's goal.

---

[4] http://www.loa.istc.cnr.it/dolce/overview.html.
[5] For examples of complexity management techniques based on foundational ontologies one could refer to [10].

Imagine we have developed a domain ontology that includes the following piece of information: "Spouses can be exclusively married to one another".

– What makes this relation true for a given pair (if true): this would be done by explicitly representing the *truthmaker* of that relation, i.e., the Marriage relator [11]. In other words, e.g., John is married to Mary iff there is a particular Marriage relator binding them. Moreover, the model could explain that 'spouse' is a *role* played by adult individuals of the *kind* Person when bound by a marriage relator.
– What is the semantics of that relation and what does it entail: by representing a relator as a bundle of legal dispositions (e.g., commitments, claims, liabilities, powers, etc.), we would be able to explain exactly what it means to be 'married with', i.e., what it means in terms of its (legal) consequences to be a spouse in that context.
– Improving comprehensibility by abstracting legal dispositions, as well as a marriage relator and the events in its life cycle, into a direct (but further explainable) 'married with' relation with the proper cardinality constraints.
– Exemplify that 'John and Mary are married' in a given timespan as an example from the data.
– Demonstrate that 'John cannot also be married to Clara in a timespan intersecting that one', because the notion of marriage in that model is of a monogamous one (as a negative example or contrastive explanation).

Ideally, these explanations should be provided in different forms. Actually, algorithms for producing such explanations already exist, but techniques for complexity management of ontologies, e.g., modularization and abstraction [10], can be considered as explanation techniques if and only if they take into account the user's goals.

5   Final Considerations

One of the primary functions of explanation is to facilitate learning [16]. Before reusing an already existing domain ontology one needs to understand it, thus, making ontology explanation an important feature of any ontology management system. However, ontology comprehension is a complex process that happens in the human mind, and which needs a proper support from the software side. Unfortunately, when talking about ontology explanation researches are mostly focused on considering different forms, instead of providing connections to different levels. Currently, there are already existing algorithms that can provide support to some of the explanation requirements outlined here, but to the best of our knowledge, there is no tool that would cover all required functionality. Also, it would be interesting to understand whether such a tool will be able to have a positive effect on ontology reuse.

References

1. Baron, J.: Forms of explanation and why they may matter. Cogn. Res. Principles and Implications **3**(1), 1–9 (2018). https://doi.org/10.1186/s41235-018-0143-2
2. Braga, B.F.B., Almeida, J.P.A.: Modeling stories for conceptual model assessment. In: Jeusfeld, M., Karlapalem, K. (eds.) ER 2015. LNCS, vol. 9382, pp. 293–303. Springer, Cham (2015). https://doi.org/10.1007/978-3-319-25747-1_29
3. Braga, B.F.B., Almeida, J.P.A., Guizzardi, G., Benevides, A.B.: Transforming OntoUML into Alloy: towards conceptual model validation using a lightweight formal method. Innovations Syst. Softw. Eng. **6**(1), 55–63 (2010). https://doi.org/10.1007/s11334-009-0120-5
4. Cao, T.Y.: Ontology and scientific explanation. In: Cornwell, J. (ed.) Explanations: Styles of Explanation in Science, pp. 173–196. Oxford University Press (2004)
5. Fernández-López, M., et al.: Why are ontologies not reused across the same domain? J. of Web Semantics **57**, 100492 (2019). https://doi.org/10.1016/j.websem.2018.12.010
6. Fonseca, F., Martin, J.: Learning the differences between ontologies and conceptual schemas through ontology-driven information systems. J. of the Association for Information Systems 8(2), 129–142 (2007). https://doi.org/10.17705/1jais.00114
7. Friedman, M.: Explanation and scientific understanding. The J. of Philosophy **71**(1), 5–19 (1974). https://doi.org/10.2307/2024924
8. Garcia, A.C.B., Vivacqua, A.S.: Grounding knowledge acquisition with ontology explanation:a case study. J. of Web Semantics **57**, 100487 (2019). https://doi.org/10.1016/j.websem.2018.12.005
9. Guarino, N., Oberle, D., Staab, S.: What Is an Ontology? In: Staab, S., Studer, R. (eds.) Handbook on Ontologies. IHIS, pp. 1–17. Springer, Heidelberg (2009). https://doi.org/10.1007/978-3-540-92673-3_0
10. Guizzardi, G., Prince Sales, T., Almeida, J.P.A., Poels, G.: Automated conceptual model clustering: a relator-centric approach. Software and Systems Modeling pp. 1–25 (2021). https://doi.org/10.1007/s10270-021-00919-5
11. Guizzardi, G., et al.: UFO: Unified Foundational Ontology. Appl. Ontol. **17**(1), 167–210 (2022). https://doi.org/10.3233/AO-210256
12. Hempel, C.G.: Aspects of scientific explanation. In: Aspects of scientific explanation and other essays in the philosophy of science, pp. 331–496. Free press Collier-Macmillan, 2nd printed edn. (1966)
13. Horne, Z., Muradoglu, M., Cimpian, A.: Explanation as a cognitive process. Trends Cogn. Sci. **23**(3), 187–199 (2019). https://doi.org/10.1016/j.tics.2018.12.004
14. Horridge, M., Parsia, B., Sattler, U.: Explanation of OWL entailments in Protege 4. In: Proc. of the Poster and Demonstration Session at the 7th Int. Semantic Web Conf. (ISWC). CEUR Workshop Proceedings, vol. 401. CEUR-WS.org (2008), http://ceur-ws.org/Vol-401/iswc2008pd_submission_47.pdf
15. Lipton, P.: What good is an explanation? In: Cornwell, J. (ed.) Explanations: Styles of Explanation in Science, pp. 1–22. Oxford University Press (2004)
16. Miller, T.: Explanation in artificial intelligence: Insights from the social sciences. Artif. Intell. **267**, 1–38 (2019). https://doi.org/10.1016/j.artint.2018.07.007
17. Overton, J.A.: Explanation in Science. Ph.D. thesis, The University of Western Ontario (2012), https://ir.lib.uwo.ca/etd/594
18. Peñaloza, R., Sertkaya, B.: Understanding the complexity of axiom pinpointing in lightweight description logics. Artif. Intell. **250**, 80–104 (2017). https://doi.org/10.1016/j.artint.2017.06.002

19. Salmon, W.C.: Four Decades of Scientific Explanation. University of Pittsburgh Press (1990), http://www.jstor.org/stable/j.ctt5vkdm7
20. Weber, E., Bouwel, J.V., Vreese, L.D.: Scientific Explanation. SpringerBriefs in Philosophy, Springer (2013). https://doi.org/10.1007/978-94-007-6446-0

Quasi-Equivalent Concept Trade-Off in Ontology Design: Initial Considerations and Analyses

Vojtěch Svátek[✉][iD], Anna Nesterova, and Viet Bach Nguyen[iD]

Department of Information and Knowledge Engineering,
Prague University of Economics and Business, Prague, Czech Republic
{svatek,qnesa01,viet.nguyen}@vse.cz

Abstract. The problem of concept equivalence is often addressed within ontology alignment. A similar problem is however encountered in ontology design: the decision whether to express multiple semantically close informal concepts as one or more formal classes, for which we coin the term *concept quasi-equivalence trade-off*. We outline its formal framework as well as an initial set of decision-making criteria. We also tried to collect traces of the trade-off from two sources: the LOV vocabulary catalog and ontology design experts addressed through a questionnaire. Finally, we discuss possible modalities of a software support.

Keywords: Ontology design · Concept equivalence · Ontology merging

1 Introduction

Concept *merging* is, in semantic web realms, associated with *ontology alignment* [2], which aims to find equivalence or subsumption links between classes from *pre-existing* ontologies. Ontology alignment techniques are also usually executed for the whole ontologies *in bulk*, whether *automatically* or (less often) *interactively*, relying on the matching of entity name strings, structural patterns and instance pools. The main purpose is to achieve interoperability of data (or document) sets described by independently developed ontologies. The existence of such data sets mandates the *soft merging* of classes, whose instance bases become bi- or unidirectionally subsumed but the classes themselves are kept.

A less investigated concept merging scenario can be however identified in the process of designing a *new ontology*. On several occasions, its designer/s may consider pairs (or, generally, n-tuples) of concepts whose semantics is very close, and decide whether to merge them or keep as separate; 'quasi-equivalent'

Supported by the IGA VŠE project № 56/2021, and by the COST Action NexusLinguarum – "European network for Web-centered linguistic data science" (CA18209), supported by COST (European Cooperation in Science and Technology), www.cost.eu. We are also grateful to the providers of example cases in Section 4: Jorge Gracia, Fahad Khan, Chris Mungall and Kateřina Haniková.

O. Corcho et al. (Eds.): EKAW 2022, LNAI 13514, pp. 209–216, 2022.
https://doi.org/10.1007/978-3-031-17105-5_16

concepts may for example be identified by cross-checking verbally expressed *competency questions*. While one or more of these informal concepts may already be expressed by a class in a pre-existing ontology, the goal is not to align existing ontologies but to reach a fine-grained modeling decision for the new ontology. The result of the decision can be not only soft merging (resulting in set-theoretically linked classes), but also a hard merging (a single class, possibly reused from an external ontology), or, on the other hand, the preservation of concepts in the form of separate classes (but, most likely, linked by some non-set-theoretical property). There may be arguments both for the merging and for the separation of the concepts. From now on, we will call this situation as *quasi-equivalent concept* (QuEC) *trade-off*. We hypothesize that abstracting elements of the rationale used in this trade-off, expressing them as guidelines, and, eventually, transforming to software support, could possibly make the life easier for OE novices.

The short paper aims to serve as an initial exploration of the quasi-equivalent concept trade-off. In Sect. 2 we formulate and exemplify the QuEC trade-off, outline an initial set of criteria that may support its resolution, and hypothesize about the visible signs of such a process in existing ontologies. In Sect. 3 we consequently analyze a collection of ontologies with respect to the presence of links considered as such signs. In Sect. 4 we provide real examples of the QuEC trade-off as provided by ontology design experts through a questionnaire. Finally, in Sect. 5 we discuss possible modalities of a software support for such decision making, and in Sect. 6 we wrap up the paper. More details about the research carried out can be found in a thesis [4].

2   Quasi-Equivalent Concept Problem Input/Outcome

The problem can be characterized as follows, in terms of input and outcome:

- Input: informal conceptualization (i.e., the designer's mental model) of the domain, containing, among other, two[1] input concepts, C_1 and C_2,.
- There are two 'canonical' variants (with sub-variants) of the modeling process outcome, in terms of the content of the output formal (OWL) ontology O:
  - (Merging outcome:) O contains in its signature either
    * (Hard merging:) a class c representing both C_1 and C_2
    * (Soft merging with equivalence/subsumption:) classes c_1, c_2 such that c_1 represents C_1, c_2 represents C_2, and either $c_1 \equiv c_2$, $c_1 \sqsubseteq c_2$ or $c_2 \sqsubseteq c_1$ holds in the deductive closure of the ontology
    * (Soft merging with overlap:) classes c_1, c_2, c such that c_1 represents C_1, c_2 represents C_2, and both $c_1 \sqsubseteq c$ and $c_2 \sqsubseteq c$ hold in the deductive closure of the ontology, whilst $c_1 \sqcap c_2 \sqsubseteq \emptyset$ does not.
  - (Separation outcome:) O contains in its signature classes c_1, c_2 such that c_1 represents C_1, c_2 represents C_2, and $c_1 \sqcap c_2 \sqsubseteq \emptyset$ holds in the deductive closure of the ontology; furthermore, there is a (logical or annotation) axiom $(c_1, p, c_2) \in O$ such that p is some predicate expressing the 'relatedness' of two concepts in other than set-theoretical terms.

---

[1] Variants for more than two concepts could be derived in a combinatorial manner.

Notably, real-world cases need not fully correspond to such 'canonical' structures, for example, in the separation outcome, the disjointness axiom $c_1 \sqcap c_2 \sqsubseteq \emptyset$ may not be present explicitly. The model also does not explicitly handle the setting with $\mathcal{C}_1$ and/or $\mathcal{C}_2$ already mapped on class/es from existing ontologies. Presumably, such classes would then be reused in the new ontology.

As an example, consider the design of an ontology of academic positions and grades. $\mathcal{C}_1$ could then be the concept of *Professor* as a role associated with a particular position at a university (among other, implying being a head of a group), and $\mathcal{C}_2$ the concept of *Professor* as being a grade recognized nation-wide and entitling, as such, to executing some responsibility by the law, at whatever academic institution. Both concepts however correspond to a person role requiring university education, implying the right to supervise PhD students, etc. A (soft) merging outcome could be, for example, the setting with three classes: *ProfessorByPosition*, *ProfessorByGrade*, and their common superclass *Professor*. A separation outcome, in turn, would be that of the first two classes being merely interconnected by a 'relatedness' predicate, for example:

:*ProfessorByPosition skos:closeMatch :ProfessorByGrade*

Various factors may influence the decision of the ontology designers. Among other, *merging* may be supported by the following arguments:

- *M1*: The ontology has to be kept small, for manageability/comprehensibility concerns (this only supports the hard merging).
- *M2*: Merging the concepts allows to keep all respective data instances under the same type, making the management of data easier.

On the other hand, *separation* may be supported by the following arguments:

- *S1*: Few or no plausible axioms could be formulated for the merged concept, while the separate concepts could be axiomatized more richly.
- *S2*: There are stakeholders behind each of the concepts who prefer to see it as separate (this is consistent with soft merging but not with hard merging).

In practical terms, how would the process of resolving the QuEC trade-off be manifested in an ontology – considering we can only access the content of O, and not the informal concepts $\mathcal{C}_1, \mathcal{C}_2$ (which were just in the heads of the ontology engineers) or discussions with stakeholders? Consequently to the above discussion, we can expect that the *merging outcome* would result in: (1) *equivalence* or *subclass* axioms in the ontology; (2) class definitions *poor in axioms*. Since the subclass axioms would most often truly correspond to subordination rather than to quasi-equivalence of the pre-cursor informal concepts, and the scarcity of axioms can also have numerous other reasons, the only sensible sign of *merging* seems to be the presence of *equivalence axioms*. The *separation outcome*, in turn, would result in pairs of classes being declared as *disjoint* but connected by some *linking property* expressing their relatedness.

In all, the possible (but, surely, not fully discriminative) manifestation of the quasi-equivalence tradeoff in the design of an ontology seems to be the presence of a pair of classes directly interconnected by a certain kind of axiom: equivalence, disjointness, or the assertion of a linking property.

3   LOV Link Analysis

Referring to the above considerations, we set out on analyzing, quantitatively and qualitatively, the structure of the ontologies indexed by the *Linked Open Vocabularies* (LOV) catalog,[2] starting from the presence of the three kinds of axioms (equivalence, disjointness, linking property). This analysis is still ongoing; some initial results (merely for equivalence and linking properties) follow.

Via a literature review we identified 21 candidate linking properties, of which we shortlisted four well-known ones (their approximate count in LOV ontologies, as of November 2021, is in parentheses): *rdfs:seeAlso* (7000), *owl:sameAs* (5000), *skos:exactMatch* (700) and *skos:closeMatch* (300). *owl:equivalentClass* axioms (among named classes) were even more frequent (14000).

Examples of possible (separation) results of the QuEC tradeoff are:

- *dbo:Annotation*[3] *owl:equivalentClass bibo:Note*[4]
- *cwmo:Idea*[5] *rdfs:seeAlso skos:Concept*[6];
- *swrc:PersonalName*[7] *owl:sameAs foaf:name*[8]
- *ldr:Agent*[9] *skos:exactMatch odrl:Party*[10]

All these correspond to concepts that are declared, at lexical level, as synonyms by respected (e.g., Oxford's) dictionaries. At the same time, their textual descriptions in the ontologies indicate subtle differences in their features.

4   Real-World Cases

We compiled a questionnaire on the QuEC trade-off that we advertised, throughout 2021, via direct mailing (to approx. 50 experts) and a few mailing lists, to the ontology engineering community,[11] yielding three fillings.[12] Additionally, we introduced a fourth case, which arose in an ongoing project related to a SARS-CoV-2 antigen testing knowledge graph, at our institute.

---

[2] https://lov.linkeddata.es/dataset/lov/.

[3] http://dbpedia.org/ontology/Annotation.

[4] http://purl.org/ontology/bibo/Note.

[5] http://purl.org/cwmo/#Idea.

[6] http://www.w3.org/2004/02/skos/core#Concept.

[7] http://sparql.cwrc.ca/ontologies/cwrc#PersonalName.

[8] http://xmlns.com/foaf/0.1/name.

[9] http://purl.oclc.org/NET/ldr/ns#Agent.

[10] http://www.w3.org/ns/odrl/2/Party.

[11] The questionnaire is still ready for input, at https://forms.gle/ZBXyfzXwmBC 8ymob9.

[12] The reason for this low response may be the unfamiliarity of the topic under the given framing, in combination with the Covid-19 pandemics.

4.1 Case 1: *Entry* vs. *LexicalEntry* in *OntoLex*

The concept *LexicalEntry*[13] pre-existed in the core module of the *Ontolex* ontology. When the new *lexicog* (for 'lexicography') module was being developed, a concept called *Entry*[14] was proposed for it, which considered the position of the entry in a dictionary rather than merely its linguistic features. Although the semantics of the concepts was similar, both were retained (after consultation with experts), in order to provide the 'lexicographic view' of the entry for the respective stakeholders while at the same time allowing to only use the core module when the lexicographic view is not essential. The module-internal *describes*[15] property was proposed to express the link from *Entry* to *LexicalEntry*.

4.2 Case 2: *Attestation* in *lemonBib* vs. *Citation* in *CiTO*

In the *lemonBib*[16] ontology it was deemed useful to model the notion of *Attestation*, similar to the notion of *Citation*[17] in the existing CiTO ontology. The two concepts were however identified as pertaining to different levels of description [3]. In lexicography, *attesting* some property of a word means referencing an external text in which this property is manifested by a word occurrence. According to CiTO, a citation is "a conceptual directional link from a citing entity to a cited entity, created by a human performative act of making a citation". This definition ignores the purpose of citing, which was, however, crucial for *lemonBib*; for example, a citation may refer to a word occurrence in order to attest a particular one of its senses, or its rhetorical role, which each correspond to a different attestation target (while the citation target remains the same). Therefore, the entities were kept as separate. To capture their interrelationship, a custom linking property *attestationCitation*[18] was used to connect their instances.

4.3 Case 3: *Fanconi Anemia* in *Mondo Disease Ontology*

Mondo Disease Ontology has been semi-automatically merged from multiple disease resources. One of the merged concepts is that of *Fanconi anemia*,[19] a hereditary DNA repair disorder. It had been a sub-concept of numerous concepts in the source models; these concepts mostly address a specific organ/tissue whose development is affected by the disorder, e.g., 'genetic skin disease' or 'congenital limb malformation'. The quasi-equivalence was concluded to be a true equivalence (the same disorder), while the positioning of the merged concept in 11 different branches of the ontology reflects its diverse perceived manifestations.

---

[13] http://www.w3.org/ns/lemon/ontolex#LexicalEntry.

[14] http://www.w3.org/ns/lemon/lexicog#Entry.

[15] http://www.w3.org/ns/lemon/lexicog#describes.

[16] http://lari-datasets.ilc.cnr.it/lemonBib.

[17] http://purl.org/spar/cito/Citation.

[18] http://lari-datasets.ilc.cnr.it/lemonBib#attestationCitation.

[19] http://purl.obolibrary.org/obo/MONDO_0019391.

4.4   Case 4: Notions of 'Evaluation' in the Antigen Test Ontology

In the context of developing[20] a knowledge graph on various kinds of SARS-CoV-2 antigen tests, a number of concepts are being considered for the ontological schema, some of which have the character of 'evaluation' of a test. Some 'evaluations' are, essentially, *claims* (on test sensitivity) made by manufacturers based on their proprietary sources. Some 'evaluations', in turn, are statements made by independent organizations or bodies, already having the character of *certification*. Furthermore, some of these independent evaluations are accompanied with quantitative results from either in vitro or clinical studies (again, as sensitivity figures), while some other are mere verdicts (passed/failed). Finally, the tests are also 'evaluated' with respect to their *listing* within national or EU-level lists. The publishers of the lists however do not perform any study; they merely verify the fulfillment of common criteria through existing studies. For example, a test listed in the EU Common List should reach at least a 90% sensitivity and a 97% specificity,[21] and must have been validated by at least one Member State based on a study providing details on the methodology.

The plethora of trade-offs remains yet unresolved, but the separation of 'claims' from 'certifications' appears more likely than their unification. On the other hand, the independent evaluations by authorities may deserve a common over-arching class, whether quantitative evidence is present or not. Finally, the notion of 'list' should be modeled separately from that of 'evaluation', but their instances should be connected via a domain property.

4.5   Comparison of the Cases

Two of the cases (3 and 4) are from the biomedical domain; this is unsurprising given the prominent role of this domain in knowledge/ontology engineering research. The reason why there are also two cases from linguistics/lexicography can be explained by an initial bias in choosing the direct mailing subjects.[22] As regards the criteria used to merging/splitting the quasi-equivalent concepts, apparently, in Cases 2 and 3 it was primarily their semantic 'essence' of the concepts; the same will probably hold for the ultimate decision in Case 4. In Case 1 it seems that the semantic difference might have been accommodated within one concept (*lexicog*'s *Entry*), the positioning information only being optional; the assumption of two different stakeholders groups (one requiring the richer version of the concept in *lexicog*, and one being fine with the core *Ontolex*), however lead to separation.

5   Software Support Considerations

Starting from the premise that a criterion in the QuEC trade-off is the proportion of axioms in/valid for both quasi-equivalent concepts, the interplay between

---

[20] Starting from a database source behind the https://covidtesty.vse.cz/english/ portal.

[21] https://data.consilium.europa.eu/doc/document/ST-5451-2021-INIT/en/pdf, p. 8.

[22] Namely, the fact that the research is partially aligned with the Nexus Linguarum COST Action, https://nexuslinguarum.eu/.

concepts and their axioms in the ontology will be of interest. This leads us to seeking inspiration from *knowledge elicitation* techniques, such as the *personal construct theory* made popular in the 1980s through the ETS system [1]. During the process of incrementally eliciting entities and their features from the expert, the tool repeatedly asks either about features that are common or discriminate between given entities, or about new entities that differ from given entities (in some feature). With respect to our QuEC challenge, the approach might have to be extended from the level of entities to a two-level system of *concepts and their instances*, and the role of features would be played by *structured axioms* (namely, Tbox and Abox ones) instead of propositional features. The system would elicit axioms common for or distinguishing between the quasi-equivalent concepts, as well as between the instances of those concepts (potentially leading to further concept splitting). A criterion for the merging/separation would be the number/proportion of axioms that could be asserted for the chosen constellation of concepts. The process would have a dual effect: aside the conclusion on merging/separation, the axioms would be elicited.

While in the 1980s the experts were the dominant source of knowledge, in the semantic web era we pay attention to the *reuse of structured knowledge.* In the simplest scenario, this would mean that not all the axioms brought into the analysis would have to be elicited from the user but would rather be picked up from existing ontologies or even inductively learned from knowledge graphs.

Finally, *textual resources* should be consulted. A focused version of concept description learning [5], where the axioms would be specifically sought for the chosen quasi-equivalent concepts (with the user serving as oracle, assigning them to either one or the other), might be applied.

6 Conclusions

We have presented the assumption that ontology engineers (frequently, or at least occasionally) encounter the quasi-equivalent concept trade-off, and outlined the principles that may govern the decision making in such cases. The empirical evidence collected from both existing (LOV) ontologies and experts addressed via a questionnaire is so far rather limited. While we also provide initial considerations on what kind of software support could alleviate the described challenge, further empirical research would probably be needed first in order to ascertain the cost/benefit ratio of developing such a support.

References

1. Boose, J.H.: Personal construct theory and the transfer of human expertise. In: IAAA 1984, pp. 27–33 (1984)
2. Euzenat, J., Shvaiko, P.: Ontology Matching, 2nd edn. Springer, Heidelberg (2013). https://doi.org/10.1007/978-3-642-38721-0
3. Khan, A.F., Boschetti, F.: Towards a representation of citations in linked data lexical resources. In: Euralex 2018 (2018)

4. Nesterova, A.: Management of Quasi-equivalent concepts in ontologies. MSc. thesis. Prague University of Economics and Business (2021). https://insis.vse.cz/zp/index.pl?podrobnosti_zp=75522
5. Petrucci, G.: Learning to learn concept descriptions. Ph.D. thesis, University of Trento, Italy (2018)

Author Index

Printed in the United States
by Baker & Taylor Publisher Services

Printed in the United States
by Baker & Taylor Publisher Services